T0221019

3D SCANNING FOR ADVANCED MANUFACTURING, DESIGN, AND CONSTRUCTION

3D SCANNING FOR ADVANCED MANUFACTURING, DESIGN, AND CONSTRUCTION

GARY C. CONFALONE

JOHN SMITS

THOMAS KINNARE

East Coast Metrology
Topsfield, MA, USA

WILEY

Library of Congress Cataloging-in-Publication Data
Names: Confalone, Gary C., author. | Smits, John, author. | Kinnare, Thomas, author.
Title: 3D scanning for advanced manufacturing, Design, and Construction / Gary C. Confalone, John Smits, and Thomas Kinnare.
Description: Hoboken, New Jersey : John Wiley & Sons, [2023] | Includes bibliographical references and index.
Identifiers: LCCN 2022051307 (print) | LCCN 2022051308 (ebook) | ISBN 9781119758518 (hardback) | ISBN 9781119758556 (pdf) | ISBN 9781119758563 (epub) | ISBN 9781119758532 (ebook)
Subjects: LCSH: Scanning systems. | Manufacturing processes. | Power electronics.
Classification: LCC TK7882.S3 C664 2023 (print) | LCC TK7882.S3 (ebook) | DDC 621.381/044--dc23/ eng/20221109
LC record available at https://lccn.loc.gov/2022051307
LC ebook record available at https://lccn.loc.gov/2022051308

Cover image: © Gorodenkoff/Shutterstock
Cover design: Wiley

Set in size of 9/13 and STIX Two Text by Integra Software Services Pvt. Ltd, Pondicherry, India

This book is dedicated to our colleagues and coworkers who are always by our side to help and provide their expertise without hesitation. We would also like to extend our gratitude to our clients and vendors for providing advice, guidance, and material content for this book, but mostly for their unparalleled benevolence and friendships throughout the years. And finally, we especially want to thank our beloved wives and children for their patience and encouragement during the many evenings and weekends spent working on this text.

Contents

Preface

For many years additive manufacturing and 3D scanning were considered a futuristic science seen only in Hollywood movies and television shows. This technology has now become a reality which has found its way into mainstream society.

Today, with relative ease, folks in this field can explain their vocation in a manner that others can recognize and appreciate. The career opportunities in this profession are wide and varying as technology expands the scope of these revolutionary applications through multiple industries.

3D Scanning for Advanced Manufacturing, Design, and Construction is a result of many years spent in the field and in the lab throughout the formative development of this science. The authors combine to bring over 100 years of experience as they worked in varying facets of this field, and the manner in which it was applied to their area of expertise. Aerospace, manufacturing, industrial research and development, architecture, engineering, and construction all use this technology to advance the improvement of their respective trades. Each of these business segments and many more are explored with real-world examples. To that we credit our coworkers, clients, and vendors for their support and advice throughout the creation of this text. Without their experience and diverse knowledge, the culmination of work in the following pages would not exist.

How to Use This Book

This book, written for the student, technician, designer, engineer, manager, and other manufacturing practitioners interested in acquiring and applying 3D scanning technologies and processes within their organization, aims to provide a broad foundation for ongoing learning. It covers all aspects of 3D scanning from the history of measurement, survey, and metrology to the practical applications of 3D scanning as it is used for data collection, analysis, and reverse engineering. We will explore applications ranging from radar bore-sighting to architectural engineering and construction with practical, useful, real-world case studies. The case studies further illustrate 3D scanning technology and demonstrate methodology through samples of varying applications and solutions without the burden of detailed theories.

Chapter 1 is an overview of the history of metrology, the science of measurement. From ancient times to modern day, methods and equipment are described to give the reader insight into the development of measuring methods, from handheld sticks to modern-day laser equipment.

Chapter 2 introduces the basics of laser scanning technology, beginning with a discussion of lasers, how they are created, what types exist, and the various classifications of laser instruments. The development of equipment using lasers for measurement will be reviewed, explaining the basic methodologies employed for scanning.

In Chapter 3 all of the varying instruments used to collect data are described. Laser trackers, laser scanners, and LIDAR systems are a few of the instruments to be reviewed for function, purpose, and practical applications.

Chapter 4 begins with a review of the software needed for processing scan data, how the software functions, and what are key capabilities to look for when comparing software. The chapter then explores point clouds and polygonal mesh files. What they are along with best practices for obtaining and processing the data. Data process techniques discussed include cloud and feature registration, along with polygonal meshing and cross sectioning for both engineering and architectural modeling as well as CAD to part analysis.

Chapter 5 explains the math behind data processing, analysis, and reporting. Critical to the application of data collection is the way it is reported to the end user. One must consider scaling and uncertainty to provide accurate and reliable data. Most important is the ability to report this data in a real-world coordinate system that makes sense to the designer as well as the end user.

Chapter 6 presents the concept of reverse engineering as used within the manufacturing and consumer industries. The varied uses of reverse engineering methods are described along with the use of 2D and 3D CAD modeling techniques unique to industrial metrology. Case studies are presented, providing the reader with real-life examples of the applications of laser measurement to modern-day projects.

Chapter 7 presents reverse engineering as used within the AEC (architectural, engineering, and construction) industry. Readers are led through the advantages scanning provides for modern-day design and construction, along with various applications of this useful technology, including clash detection, forensics, and BIM modeling. Case studies illustrate the use of multiple pieces of laser scanning equipment and the diversity of deliverables used by actual projects to provide timely and cost-saving solutions.

Chapter 8 is a brief analysis of emerging trends within the laser scanning industry, and what can be foreseen for future development for software, equipment, and applications.

Chapter 9 provides a number of resources and references for the reader, with listings of professional organizations, equipment, and software manufacturers. Universities and colleges that offer courses and degrees in metrology and the use of 3D laser scanning technology are given.

A Glossary of Terms is at the end of the book, a valuable reference for the reader to understand and define the many specialized terms and acronyms used in the metrology industry.

Navigating this book, you will be part of the journey in analyzing opportunities for the application of 3D scanning in your business or industry. Begin by identifying a problem or opportunity for which you would like to explore a scanning solution. Work through the major decision steps and post-processing needed to get to the desired end result, such as a point cloud or a solid model. The case studies will explain how others have addressed challenges and decisions to move 3D scanning forward in their business workflows. By the end of the book, you will be prepared to make a case for implementing 3D scanning technology into your next project.

Chapter 1
History of Metrology

The regulation of weights and measures is necessary for science, industry, health care, and commerce. The importance of establishing uniform national standards was demonstrated by the drafters of the US Constitution, who gave Congress in Article 1, Section 8, the power to fix the Standard of Weights and Measures. "Weights and Measures," said John Quincy Adams in 1821, "may be ranked among the necessaries of life to every individual of human society."[1]

> Weights and measures may be ranked among the necessaries of life, to every individual of human society. They enter into the economical arrangements and daily concerns of every family. They are necessary to every occupation of human industry; to the distribution and security of every species of property; to every transaction of trade and commerce; to the labours of the husbandman; to the ingenuity of the artificer; the studies of the philosopher; to the researches of the antiquarian; to the navigation of the mariner, and the marches of the soldier; to all the exchanges of peace, and all the operations of war. The knowledge of them, as in established use, is among the first elements of education, and is often learnt by those who learn nothing else, not even to read and write. This knowledge is riveted in the memory by the habitual application of it to the employments of men throughout life. (John Quincy Adams, *Report to Congress*, 1821)

---

[1] Bucher (2004). *The Metrology Handbook*. https://ebookcentral.proquest.com/lib/umaine/reader. action?docID=3002524, p. 1–9.

3D Scanning for Advanced Manufacturing, Design, and Construction, First Edition.
Gary C. Confalone, John Smits, and Thomas Kinnare
© 2023 John Wiley & Sons, Inc. Published 2023 by John Wiley & Sons, Inc.

1.1 INTRODUCTION

To understand 3D laser scanning technology, a person must have an understanding of metrology. Metrology is defined as the science of measurement. It is the language that engineers use to communicate to manufacturers.

When studying science, technology, engineering, and math, you will use units and the universal language of metrology which was developed thousands of years ago and continues to evolve today. Often the parameters of these units are referred to as geometric dimensioning and tolerancing (GD&T). This language consists of formulas, numbers, and symbols that when interpreted correctly can yield the most magnificent of outcomes. It is the language of technical professionals in manufacturing and construction, or the language the designer uses to describe what he wants to the builder and end user. And once you learn this language and become indoctrinated into this world, you will see things in a different light. I often tell students to look at my coffee cup and tell me what you see. I go on to explain that everything in this classroom was designed by an engineer using a blueprint or a formula. The blueprint will dimension the cup; and the formula will define the contents. Both the blueprint and the formula use metrology to make sure they are built to the design specifications or the proper recipe.

So, you may ask, what does this have to do with 3D laser scanning. This book will focus on the applications of laser scanners as they are used to measure and reproduce 3D results. Often the reproduction of these results is referred to as Reverse Engineering or As-Built documentation. Both topics will be explored in the subsequent text.

1.2 THE HISTORY OF METROLOGY

Archeologists believe that measurement standards have been with us for over 6,000 years and probably longer. With the adoption of agriculture to what was once a nomadic species, humans needed a way to measure their land and crop yields to communicate fair trade to others. But measurement was not only limited to the agricultural industry, free trade also created a need for measurement in all facets of life in a growing industrial world market. This of course resulted in a wide variety of measurement systems being developed throughout the world.

It was not until 1875 that engineers and scientists began to establish an internationally recognized system at the "Metre Convention" held in Paris. At the Metric Convention of 1875, as it is called in English, a treaty was signed between 17 countries including the US to establish the international Bureau of Weights and Measures, which would work to standardize the four basic measurement standards: mass (weight), distance or length, area, and volume. [Today standardization also includes temperature, pressure, luminosity, and electric current.]

1.3 THE INTERNATIONAL SYSTEM OF UNITS (SI)

In 3D scanning and surveying, length, angle, area, and volume are the primary units of measure. The two systems used for specifying these units of measure in the world today are the Metric System and the British Imperial System.

The Metric System was developed in France in late eighteenth century and is maintained by the General Conference on Weights and Measures (GCWM). Since the metric system is almost universally used today, it is often referred to as the International System of Units and abbreviated SI (Système International d'Unités).

Units in the British Imperial System (BIS or IS) are derived from the English System of units which is rooted in historical units from both Roman and Anglo-Saxon units. To make things more complicated, the *US Customary Measurement System* is a system based on the English System which was the measurement system used in Britain prior to the introduction of the British Imperial System in 1826.

I know all of this can be confusing and you can only imagine what it was like hundreds of years ago before the internet was able to do a unit conversion. The differences between the US and British systems are only as they relate to volume and will not impact the discussions in this text. Our focus here will primarily be concerned with dimensional metrology.

A Message from the President to the Senate of the United State:

> I transmit to the Senate for consideration, with a view to ratification, a metric convention between the United States and certain foreign governments, signed at Paris, on the 20th of May 1875, by Mr. E. B. Washburne, the minister of the United States at that capital, acting on behalf of this government, and by the representatives acting on behalf of the foreign powers therein mentioned. (Washington, *March* 10, 1876. Ulysses S. Grant)

1.4 THE HISTORY OF THE METRIC SYSTEM

With over 700 recorded units of measure in France, a movement was made after the French Revolution where engineers looked for standards that were based on pure natural occurring physics. Strangely enough around 1790, the Metre was introduced as one ten-millionth of the shortest distance from the North Pole to the equator (Quarter Meridian) passing through Paris. To define this, the measurements and construction of the standards were entrusted to the *Institute of France* and international representatives who served as deputies of this commission. *Jean-Baptiste Joseph Delambre* and *Pierre-Francois Meçhain* then set forth to identify the meter by making geodetic and astronomical measurements along the meridian from Dunkirk to Barcelona. This took 7 years of extensive survey triangulation work to complete. From these results there was constructed a one-meter bar of platinum whose length was measured between its two ends, and it became known as the "Meter of Archives."

Table 1.1 Metric system – length units of measure

Base Unit	Symbol	Meter Equivalent
Kilometer	km	10^3
Meter	m	1 m
Decimeter	dm	10^{-1} m
Centimeter	cm	10^{-2} m
Millimeter	mm	10^{-3} m
Micrometer	μm	10^{-6} m
Nanometer	nm	10^{-9} m

By 1795 the Decimal Metric system evolved to recognize the Metre as the standard unit of measure. This simplistic system soon became a weights and measures law throughout the globe noting also that a cube having sides of length equal to one-tenth of a meter was to be the unit of capacity, the liter, and the mass of a volume of pure water equal to a cube of one-tenth of a meter at the temperature of melting ice was to be the unit of mass, the kilogram.

Referring to the original standards became difficult over the years so to be more practical, the platinum bar that was held in Paris to define this metre (meter) was replaced 100 years later in 1889 by the International Geodetic Association with 30 platinum–iridium bars that were distributed throughout the world. It was not until 1960 that a new definition was derived using the physical properties of light. The spectral emission of Krypton-86 radiated light at 606 nanometers (orange) became the new international length standard from 1960 to 1983.

Today the meter is defined by research performed by the National Institute for Standards and Technology (NIST) as the length of light travel in 1/299,792,458 of a second in a vacuum. Table 1.1 lists the SI units of measure commonly used to define length.

1.5 THE HISTORY OF THE BRITISH IMPERIAL SYSTEM (IS)

So as not to be outdone by the French, the Imperial System of units was established by the British Weights and Measures Act of 1824. The imperial units were preceded by the Winchester Standards which were in place from 1588 to 1825. Derived from hundreds of Roman, Celtic, and Anglo-Saxon units, the British Imperial System was the primary measurement standard until the UK joined the European Economic Community in the 1970s. However, some imperial units are still in use today. One of those being the pint which is a very important standard frequently used by metrologists in Britain, the US, and throughout the world!

Because this book is focusing on 3D scanning, we will stick to the topic of dimensional metrology. Below we will walk you through the most common units of measurement as they relate to length and give a brief history of their origin. Most of the IS units will make sense after learning the history or the etymology of the unit identifier. These include units such as the furlong, the mile, and the yard. Other units like the hand, foot, or pace have origins that are a bit more self-explanatory.

1.5.1 Inch

There are many historical accounts of how the inch was derived. Earliest would be the Anglo-Saxon definition of the inch as being the length of three grains of barley placed end to end. Later in history, King David I of Scotland and his court of Weights and Measures defined the inch as the width of an average man's thumb, measured at the base of his thumbnail.

1.5.2 Thou, Mil, and Tenth

When working in the field of metrology one must learn to adapt to the various industries and national systems under discussion. With the growing popularity of the metric system here in the United States, I often see people refer to the millimeter as a "mil" when in fact the "mil" is an abbreviation that refers to 0.001 or one-thousandth of an inch. Mil is derived from the Latin word Mille meaning one thousand. The mil is interchangeable with the other term for this measurement referred to as the "thou." Thus, one thousand thou equal one inch; and one thousand mils also is equal to one inch.

Much like the ambiguous challenges children face when spelling words in the English language, metrology will also become second nature after time but should not be taken for granted. For the students and readers not yet fully immersed in the manufacturing industry I believe it is worth building on the above and not assume that all abbreviations are clear. In discussions among engineers and machinists you will always hear the terms, thou, mil and tenths when referencing precision tolerances. We have determined above that the terms thou and mil are synonymous and refer to 0.001 inch. If you are not included in these social circles and hear the term "tenth" one may naturally assume it is indicating one-tenth of an inch. After all that makes perfect sense. But what it is actually defining is one-tenth of one-thousandth of an inch or 0.0001 inch. We won't get too bogged down by these facts, but I believe this was simply worth noting to students of the industry and others new to the field because we often see this confusion with young engineers.

1.5.3 Hand

Still in use today is the ancient tradition of using the "hand" to measure the height of horses. Horses from the ground to the top of the shoulder (withers). The unit was originally defined as the breadth of the palm including the thumb. A statute of King Henry VIII of England established the hand at four inches. Therefore, a horse that measures 15 hands would be 60 inches tall at the shoulder (Figure 1.1).

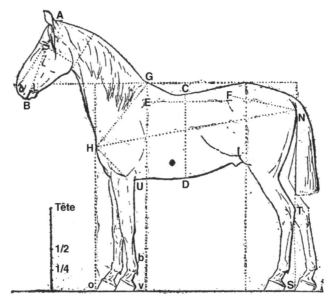

Figure 1.1 Equine metrology

1.5.4 Foot

Like the inch, there are many rumored origins of the foot as a length standard. The first recount was said to be around 2575 BC where it is speculated that the Greeks and Romans adopted the foot standard from the Egyptians. In all accounts, and like so many other units based on the human body, it is assumed that the origin of the foot standard was the length of a man's foot (Figure 1.2).

British history states that King Henry I decided his foot would be a standard for length, as pacing out length measurements was a very common practice in land surveying. This foot measurement standard was to be a booted foot rather than a bare foot. This was probably the case because folks taking these measurements or paces were donning boots while performing these measurement tasks.

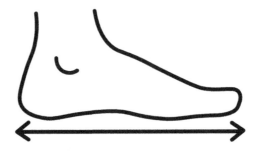

Figure 1.2 Foot length standard

1.5.5 Cubit

The earliest recognized standard for measurement is the Egyptian Cubit. This standard is based on the forearm length of the Pharaoh reigning around 3000 BC. One can easily see how this could be an issue as Pharaohs will change from one generation to the next. The cubit therefore had a range of approximately 43–53 centimeters throughout history. The flood level of the Nile circa 3000 BC was defined as 6 cubits and 1 palm; where 1 palm is equal to 4 fingers and 6 palms is equal to a cubit.

1.5.6 Yard

The first yard followed suit with most of the other standards in that it was a reference to the human body but in many various ways. It is documented that this standard was based on the single stride of a man; the breadth of a Saxon's chest; or the length of a man's girdle (belt). Yet others believe it was originally a cubic dimension, or a unit of volume. Whichever it may have been at the time, not unlike the foot and the inch, there were many variations in this standard. For this reason, many centuries of measurement uncertainty caused much disagreement in trade and contentious land disputes.

During the reign of Edgar the Peaceful (959–963 AD) he dictated that a bar he designated "the yardstick" held at the Cathedral of Winchester should be observed as the standard throughout his Realm.

And again, with a new King comes a new measurement definition. Like the foot, in the twelfth century, King Henry I redefined the yard as the distance from his nose to the thumb of his outstretched arm. Oddly as it may seem, Henry was only off by one-hundredth of an inch from today's standard!

1.5.7 Chain

Early surveyors often used a chain as a useful tool in their measurement process. Many of these chains tended to vary in length. Along came Edmund Gunter (1581–1626), a well-known mathematician and astronomer in the day. Gunter made many important scientific discoveries, one of those being his process of triangulation to survey and plot land area. He employed a chain that would allow him to take linear measurements between topographical features in the land to plot and calculate the area. For this the Gunter Chain, as it was called, became known as a standard of length and was equal to 66 feet.

1.5.8 Rod

Edmund Genter who we just spoke of, also standardized the rod as a unit of measure in 1607. A rod, also called a perch or pole, was defined as one-quarter of a chain (16½ feet) (Figure 1.3).

Figure 1.3 Determination of the rod, using the length of the left foot of 16 randomly chosen people coming from church service
Source: Wikipedia. Woodcut published in the book *Geometrei* by Jakob Köbel (Frankfurt, c. 1535). WIKI.

Chains and rods were used for many centuries as land deal transactions were rapidly growing in North America and Australia. Our most famous surveyor George Washington used a rod which was fashioned out of dimensionally stable and somewhat flexible wrought iron. These rods were portable and easy to manipulate through the forest and bushes by a single man surveyor.

1.5.9 Furlong

Another old English unit of length was the furlong. This unit was simply derived from the average length of a plowed furrow ("furrow-long," or furrow). The length was determined by the amount of time a team of oxen could plow before resting. This was a sensible length for farmers that later evolved into the acre. A standard furrow is 660 feet (220 yards) long or ⅛ mile.

1.5.10 Fathom

The fathom is a measure of length that is equivalent to 72 inches (6 feet) and was primarily used by maritime navigators to measure the depth of water.

Its origin is believed to come from the English word "faethm" which means outstretched arms. This is the longest unit originating from the human body comparison as it is defined by the distance from the middle fingertip of the right hand to the middle fingertip of the left hand. This is also the length at which a man could extend his arms while measuring rope which was very suitable.

Although there are many interpretations of its origin, the most popular among historians for the term *"to deep six"* something, comes from its reference to a burial at sea where it was required as late as the early 1900s to perform a burial in at least 6 fathoms of water to assure it will not wash ashore. Today in US the depth has increased to 100 fathoms (600 feet) of ocean water that is 3 nautical miles from shore.

1.5.11   Acre

It is worth noting the history of the Acre for the benefit of the Engineering and Surveying students and readers. Tracing back to Middle England agriculture, the acre was thought to be defined by the quantity of land that could be tilled with a wooden plow and yoke pulled by oxen in the span of one day (Figure 1.4). When you plow, you make a trench in the soil called a furrow or furlong as previously noted. One pass across a field would leave a trench one "furrow long." The number of furrows cut in one day was adopted by England to be the standard known as the acre. It was later given a new and more distinct definition as one-tenth of a furlong by the Anglo-Saxons, which was approximately 660 feet in length. From this we can calculate 40 × 4 rods is roughly equivalent to 660 × 66 feet, which is where we obtain the area of 43,560 square feet used to define an acre today.

Today the acre is most commonly used by the United States and United Kingdom to describe large spans of land such as farmland, forests, and building property.

In countries that employ the Metric System of units, area will be measured in square-meters (m^2). It is therefore important to be able to convert from the US customary and Imperial System to the more commonly used International System of units (SI). Below are common conversions of acres, miles, and meters units of measure:

1 acre	43,560 square feet *(4046.856 m^2)*
1 square mile	640 acres

Figure 1.4 An acre was determined by the amount of furrows plowed in 1 day

In Chapter 9 a more comprehensive table of US and metric measurements and units is provided for the reader's reference.

1.5.12 Pace

In ancient Rome, the pace (*passus*) was measured from the heel of one foot to the heel of the same foot when it next touched the ground as done in a normal step when walking. This is a convenient unit for measuring walking distances such as the mile as described below. A standard pace is 5 Roman feet long.

1.5.13 Mile

As interesting as the Egyptian cubit is another historic length standard known as The Roman Mile deriving from the Roman mille passus, or "thousand paces." This unit of measure was defined as a count of 1,000 paces of which each pace consisted of 2 steps or 5 Roman feet. Thus, the Roman Mile was equal to 5,000 Roman feet. Roman Armies would literally put stakes in the ground marking out each mile as they marched. However, measurements taken this way introduced variables of uncertainty beyond physical attributes. The conditioning of the soldiers and the weather that they marched in would impact the stride and therefore the length of the mile.

Now that you know how many of the old measurement units that have stood the test of time were derived, Table 1.2 below illustrates today's definition of the seven (7) base units and how they are defined in most of the modern world.

Table 1.2 SI-based units and standards

Base Measure	Name	Symbol	Definition
Time	second	s	The second is defined by taking the fixed value of the cesium frequency $\Delta \nu$Cs, the unperturbed ground-state hyperfine transition frequency of the cesium-133 atom, to be 9,192,631,770 when expressed in the unit Hz, which is equal to s^{-1}.
Length	metre	m	The length of the path traveled by light in a vacuum during a time interval of 1/299792458 of a second.
Mass	kilogram	kg	Defined (since 2019) by taking the value of the Planck constant, h, to be $6.62607015 \times 10^{-34}$ when expressed in the unit J s, which is equal to $kg \, m^2 \, s^{-1}$.

(Continued)

Table 1.2 *(Continued)*

Base Measure	Name	Symbol	Definition
Electric current	ampere	A	Defined (since 2019) by taking the value of the elementary charge, e, to be $1.602176634 \times 10^{-19}$ when expressed in the unit C, which is equal to A s.
Thermodynamic temperature	kelvin	K	Defined (as of 2019) by taking the value of the Boltzmann constant, k, to be 1.380649×10^{-23} when expressed in the unit J K^{-1}, which is equal to kg m^2 s^{-2} K^{-1}.
Amount of substance	mole	mol	Contains (since 2019) exactly $6.02214076 \times 10^{23}$ elementary entities. This number is the fixed numerical value of the Avogadro constant, N_A, when expressed in the unit mol^{-1}.
Luminous intensity	candela	cd	The luminous intensity, in a given direction, of a source emitting monochromatic radiation of a frequency of $540 \times 1{,}012$ Hz with a radiant intensity in the direction of 1/683 watt per steradian.

1.5.14 Why Isn't the US Adopting the Metric System?

Many years ago, I recall learning the Metric System in grammar school. We were told that within a few years the United States would be switching to the Metric System, and it would become our primary system of measure. After all, it was the universal language and much easier to use. It was a big topic of discussion in every household in America, and I even recall *Saturday Night Live* doing a skit about the new *Metric Alphabet now called the "Decabet"* which would shorten our current alphabet to 10 letters! So, it wasn't a surprise when I graduated engineering school and joined the workforce that I was to be part of an aerospace program that was decidedly going to design an aircraft entirely using the metric system. This would make support easier as the mechanics would only need metric tools and would hopefully make the aircraft more attractive to the international market. Imagine a toolbox with one set of metric tools to service the entire fleet. All of the hardware was to be metric and all of the hydraulic and electrical fittings were to be Metric. This plan sounded great; however, there were a lot of suppliers that needed to be convinced and there really was not much in terms of spare metric hardware within the corporation. This did cause many delays because lean manufacturing practice was to have on hand only what was needed and if you stripped a thread, it was very difficult to quickly find a spare. And as far as the suppliers of electrical connectors and hydraulic lines were concerned, the cost would be prohibitive.

I will admit that the metric system made math and physics calculations much easier as an undergrad in college, and with a few exceptions it is somewhat surprising that it has taken so long to be embraced by big manufacturing. The Britannica Encyclopedia suggests the sheer size and growth rate of the United States made a total transition from the British System to the Metric System costly and inhibiting:

> The biggest reasons the U.S. hasn't adopted the metric system are simply time and money. When the Industrial Revolution began in the country, expensive manufacturing plants became a main source of American jobs and consumer products. Because the Imperial System (IS) of measurements was in place at this time, the machinery used in these factories was developed to size in IS units; all of the workers were trained to deal with IS units; and many products were made to feature IS units. Whenever the discussion of switching unit systems arose in Congress, the passage of a bill favoring the metric system was thwarted by big businesses and American citizens who didn't want to go through the time-consuming and expensive hassle of changing the country's entire infrastructure. Many also believed that the United States should keep its particular system, setting it apart from other countries and symbolizing its status as a leader rather than a follower. (Figure 1.5)

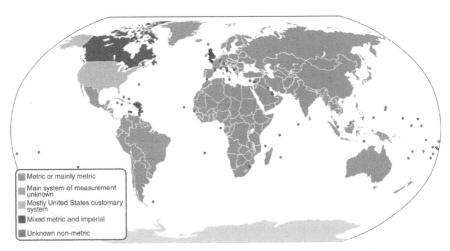

Figure 1.5 Countries using the metric, US customary and imperial systems of units as of 2019. Only the USA uses the US customary system; Liberia, Samoa, Palau, Micronesia, and Marshall Islands use unknown (i.e. it is unclear which system they use) nonmetric systems; it is impossible to clearly determine which system Myanmar uses; the UK and Canada use a mix of imperial and metric systems
Source: Goran tek-en 2020 / Wikimedia / CC BY SA 4.0.

In modern times, most have accepted a joint unit system – teaching children in school both the traditionally used IS system and the metric system that most of the rest of the world uses. This is why US measuring sticks, or rulers, often contain both inches and centimeters. Unfortunately for metrics fans, widespread acceptance of joint use also means that there likely will be no official phasing out of the IS system anytime soon. (Hogback)

1.6   EVOLUTION OF METROLOGY
1.6.1   Industrial Revolution

Throughout the years, the need for widely adopted and repeatable standards grew. Not unlike the advancements in technology today, the continued improvements and advances in science and technology of yesterday were driven through the need for better military equipment. Cannon bores, projectile diameters, and integrated machinery drove the necessity for high precision equipment early on. In the dawn of engineering, weapons and weapon systems were the most interesting and complex designs of the time. Some of the greatest engineers in history designed and built weapons that are still being manufactured to the original specifications today. This industry fostered advances in both machinery and material science. This evolution in engineering continued to expand with the development of the automobile and aerospace industry,

The design and production of sophisticated mechanisms also led to adoption of a modern metrology system, including master gages, transfer standards, and regular comparators (Figure 1.6).

In 1789 Eli Whitney won a government contract to produce 10,000 firearms having interchangeable parts. This is how the first gage block was developed. Throughout history many companies have been recognized for their advanced designs and superior products. What is not often discussed is the development of the machines to produce and inspect these products. Robbins and Lawrence is another New England

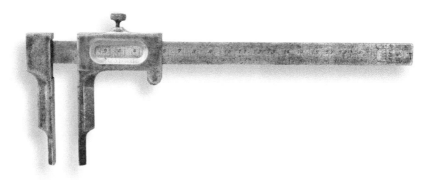

Figure 1.6 Antique vernier caliper

company that was mass producing weapons for both the US and British governments in the 1800s. Until these developments successful interchangeability between rifle parts was very rare as each rifle was hand assembled with unique individually made parts.

By the mid-1800s a skilled machinist was producing hundreds of parts with a dimensional repeatability of ± 0.002 inch. And around that same time, Joseph R. Brown invented and manufactured the first vernier scale caliper with a resolution of 0.001 inch. This was soon followed by the micrometer which was developed 10 years later in 1867. And by the year 1877, the micrometer was commonplace in the metalworking industry.

It wasn't until the mid-1950s that an inventor named Harry Ogden developed a more sophisticated electronic measurement machine that is known today as the Coordinate Measuring Machine (CMM). Ogden worked as the Chief Mechanical Engineer at the Numerical Control (NC) Division of Ferranti Ltd in Scotland, a machine tool manufacturer. The *Ferranti Inspection Machine* was capable of measuring parts that previously took hours to inspect, in a matter of minutes. The reduction of inspection time and the simplicity of these new digital machines were about to revolutionized the quality control market (Figure 1.7).

Sheffield introduced the Ferranti CMM to the United States in 1961 under the name "CORDAX," standing for Coordinate Axes. The name soon became the generic term for these machines and Sheffield was able to sell 250 of them by 1964. The first machines performed well but the opportunity to advance the technology was rising with the competition. Digital Electronic Automation (DEA) in Turin, Italy, recognized this and demonstrated its first servo-driven 3D Gantry Style Machine at the European Machine tool show in Milan in 1965. Franco Sartorio, Luigi Azzaroni, Giorgio Minuccianni, the founders of DEA also recognized the need for a more advanced probing system that was touch sensitive.

In 1970 the cofounder of Renishaw, Sir David McMurtry, invented the Touch Trigger Probe which enabled the Olympus engines used in the Concorde Aircraft to pass specific requirements. This innovative enhancement to the CMM revolutionized 3D

Figure 1.7 Ferranti coordinate measuring machine
Source: International Metrology Systems, Ltd.

measurement, opening the world up to accurate and automatic measurements of separate components and entire assemblies.

In 1973 DEA delivered the first automated CMM with probe changing capability to Caterpillar Company in Peoria, Illinois. With more advancements in machine tool integration and scanning capabilities, DEA quickly became the world's largest supplier of CMMS. By 1994, DEA became part of Brown and Sharp in one of the many consolidations to be seen in the Inspection industry.

Although CMMs transformed the industry, they were not the answer to all of the inspection tribulations. Portable CMM technology was invented in the mid-1970s when ROMER cofounder Homer Eaton, of the Eaton Leonard Corporation, filed for a patent of the innovative articulating arm system. The original concept was designed to measure bent tubes. Shortly after its success, Homer teamed up with Romain Granger to develop the "Romer Arm," a more general-purpose measurement device (Figure 1.8). These new portable CMMs, more commonly referred to as "Arms," were able to effortlessly obtain data in places that a conventional CMM could not access with accuracies better than 0.005 inch. Like the stationary CMMs, the Arms have seen many advancements in technology, and have made 3D scanning capability a common practice through the

Figure 1.8 Modern-day 7-axis articulating arm equipped with laser scanner and probe. These arms were commonly referred to as Romer Arms until August 2018 when Hexagon Manufacturing Intelligence renamed the product to the absolute arm
Source: Photo courtesy of Hexagon Manufacturing Intelligence, Inc.

introduction of a laser scanning probe adapter, which will be discussed in more detail in Chapter 3.

1.6.2   Early Surveying

One of the oldest professions in the world is that of the builder and surveyor. Records of surveying date back to ancient Egyptian days where this science was used to build the pyramids. The early surveyors used ropes, levels, and plumb bobs to perform measurement tasks. These were the first modern survey tools. An accurate representation of a citizen's plot size was as important as they were taxed on their land as we still are today.

1.6.3   Compass

One of the earliest survey instruments known to man was the Compass. It is thought that the compass was invented by the Chinese around 200 BC. Magnetized needles or lodestones were floated on a small piece of wood in a bowl of water. The needle would align north–south and indicate real-time directional indication. By 1300 AD the compass was being used to navigate the globe and took the form of the units we are familiar with today (Figure 1.9).

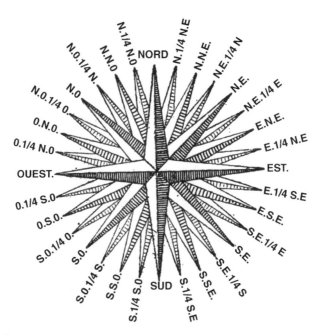

Figure 1.9 Compass

1.6.4 Diopter

An ancestor to the modern theodolite, the Diopter was developed by the Greek astronomer Hiparchus Greece around 150 BC and primarily used for land surveying and aqueduct work so popular in those times. It was a sophisticated device capable of measuring horizontal and vertical angles between two terrestrial points. In a work entitled *Diopter, Hero of Alexandria*, it is described as a portable instrument, which is a useful application of the cogwheel, screw, and water level, for taking terrestrial and astronomical measurements (Figure 1.10).

1.6.5 Transits, Theodolites, Total Stations

Much like the modern-day theodolites used for engineering and survey applications today, the original sixteenth-century theodolites had the ability to measure both vertical and horizontal angles. The most notable application of the theodolite in American history was the use by Lewis and Clark to chart and survey the Louisiana Purchase.

Following the advent of the first-generation theodolite Philadelphia manufacturer William Young invented the American Transit in 1831. The transit had the ability to perform back-sighting and doubling of angles for error reduction because it was designed to flip 180 degrees. The primary function of the transit was to project or perpetuate a straight line by turning ("transiting") the scope on its horizontal axis. The transit is

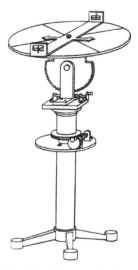

Figure 1.10 Dioptra as described by Heron of Alexandria. Wikimedia
Source: Courtesy of H. Schöne – Leipzig: B. G. Teubner, 1903Klingsor at German Wikipedia.

a lightweight useful tool on construction sites but has given way to the modern-day electronic theodolites.

The next generation theodolites are a relatively inexpensive solution to gaining accurate measurements of large parts and areas. By triangulating on a target, either using two systems or the relocation of one, the operator can determine more than just azimuth and elevation angles with arc second precision. The real innovation here lies within the data-processing software making these calculations instantaneous.

How could it get any better? In 1968, Zeiss Instruments developed the total station theodolite (TST) or simply referred to as the total station (TS) (Figure 1.11). The TS is the theodolite on steroids. It can measure both horizontal and vertical angles like the theodolite, but also has the capability of measuring distance. The distance measurement capability is accomplished by using a multifrequency carrier signal. By emitting and receiving

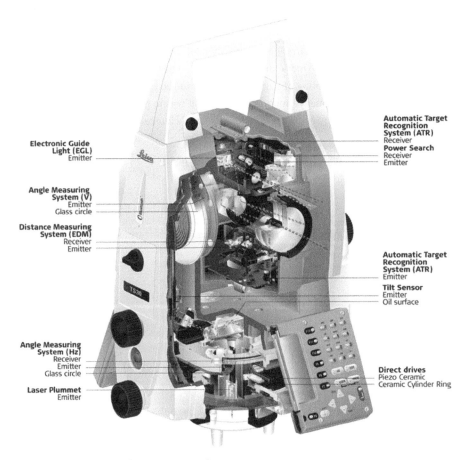

Figure 1.11 Total station cross section
Source: Photo courtesy of Hexagon Manufacturing Intelligence, Inc.

multifrequency signals, distance can be calculated by determining the number of wavelengths to the target for each signal. Most TS are capable of measuring angles with 0.5 arc-sec angular accuracy, and a distance accuracy of approximately 0.060 inch at 5,000 feet. You will often hear surveyors refer to these as "robots" because over the years, total stations have adopted motorized capability so that the operator can control the TS using a remote control.

Recently Hexagon Metrology has changed the industry by introducing the Leica Nova M60 "multi-station." This system has all of the modern capabilities of the automated TS with the added ability to perform laser scans at a rate of 30.000 points per second (Hz).

1.6.6 Laser Trackers

The introduction of the laser tracker in 1990 changed the metrology industry forever. Using a similar principle as the total station, laser tracker systems differ by being able to measure all three parameters (azimuth, elevation, distance) with extreme accuracy. The big difference being the capability of achieving distance accuracies to the center of the reflective target of 25 microns at 5 meters. Originally this distance capability lent itself to the employment of a laser interferometer within the tracking system. Laser interferometers are inherently accurate as their principle of operation is to count wavelengths. More recently, the interferometers are being replaced by a sophisticated time-of-flight laser system that is capable of achieving system accuracies close to that of the interferometer. This type of accuracy has positioned the laser tracker systems as the standard tool for high accuracy large volume metrology (Figure 1.12).

Real-time feedback from the laser tracker has made this instrument invaluable for tool building and surface scanning. More recent developments in probe technology

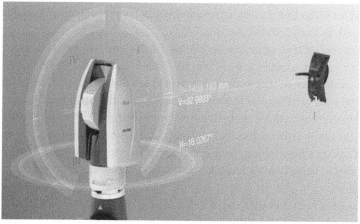

Figure 1.12 Leica Laser Tracker with 6 degree-of-freedom (6-DOF) probe technology
Source: Photo courtesy of Hexagon Manufacturing Intelligence, Inc.

have also expanded the scope of laser tracker as well as other instruments. Similar to the Portable (Arm) CMM and the stationary CMM, a laser tracker has the ability to scan by following a probe as it is swept along a surface. Recognizing that collecting date in this manner can be tedious, Hexagon developed a 3D scanning unit that when paired with the laser tracker it is capable of collecting large amounts of scan data without making contact to the surface. This tethered 3D scanner has broadened the scope of the laser tracker to be a more practical tool for reverse engineering and CAD to part comparison applications.

1.6.7   Laser Scanner

As a survey tool, laser scanners have been around since the late 1960s, but scanner capabilities were hindered by the lagging development of a processor that could handle the massive amounts of data these instruments are capable of collecting. These instruments are capable of collecting unlimited amounts of data with a typical scan exceeding 100 GB in size. Until recently, files of this magnitude would wreak havoc on most CAD packages causing them to fail. Processing time was slow making most engineers reluctant to adopt this technology early on. New developments in software and processing hardware have allowed the scanner to expand its uses in high-tech manufacturing as well as architecture, engineering, and construction.

In the mid-1980s the practicality of taking advantage of laser technology in precision measurement applications became reality. Prior to that most common were "scanners" which used lights cameras and projectors to capture the images and data. The use of a laser beam allowed for greater accuracy levels, but it did carry its own issues. A lack of computing power was one of the major issues to resolve, along with the heavy cumbersome size of the scanning devices. By the early 1990s the seminal versions of a 3D laser scanner began to take form with a system that was able to "scan" a surface and collect point cloud data. The first portable (Arm mounted) touch probe "scanner" was introduced in 1991, by Faro Technologies, Inc.

A 3D noncontact laser scanning device suitable for large-scale field work was introduced by a company named Cyra in 1996. Developed by engineer Ben Kacyra, he sought to create a device that would solve the as-built documentation issues involved in industrial plant retrofits and redesigns. It hoped to remedy two main issues: change the time-consuming methods currently used for measuring existing conditions and eliminate the costly errors associated with hand dimensioning and note taking. Within five years a functioning scanner and accompanying software had been developed and sold to Leica Geosystems, one of the world's premier surveying equipment manufacturers.

By 2004 Leica had rebranded and refined Cyra's original scan device into a portable 360-degree scanner, the HDS 3000 (Figure 1.13). Soon other companies began pro-

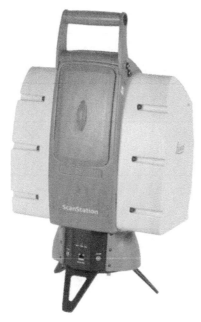

Figure 1.13 Leica HDS 3000 Laser Scanner
Source: Photo courtesy of Hexagon Manufacturing Intelligence, Inc.

ducing 3D scanners. These early units could produce 12,000 points per minute up to a 100-meter distance. By 2009 Leica created the P20, a laser scanner that produced 1 million measurements per second. Faro introduced its Photon 120 scanner, a phase-based unit capable of 976,000 points per second with a 360-degree × 320-degree field of view. Within 15 years of Cyra's initial device, Faro introduced the Focus 3D, the first small lightweight scanning device at a price point far below a typical scanner at that time. It was a disruptive moment in the evolution of laser scanning.

By 2020, within a short 25 years, scanning had become a widely used and accepted method within multiple industries. As computer memory and processing power improved, the ability, usefulness, and accuracy of laser scanners increased exponentially. Software was introduced by third-party developers that expanded the applications of 3D scan data. It had found a purposeful niche in just about every industry in the world, from manufacturing to entertainment to architecture, engineering, and construction.

The subsequent chapters will explore the various types of laser scanners and provide case studies where 3D scanning technology has changed the direction of the industry. Table 1.3 lists some of the more popular devices used today for 3D laser scanning. Each device has its own strength and weakness; some devices are more suitable for larger objects and other devices designed for higher accuracy on small, detailed parts. What-

Table 1.3 Table depicting the capability of varying units

Equipment	Noncontact	Portability	Accuracy	Speed	Cost	Automation
White/blue light scanner	✓	✓	✓✓✓	✓✓	✓✓	✓✓
Arm-mounted laser scanner	✓	✓	✓	✓✓✓	✓	
CMM-mounted laser scanner	✓		✓✓	✓	✓✓✓	✓✓✓
Handheld laser scanner	✓	✓	✓	✓✓✓	✓	
Terrestrial laser scanner	✓	✓	✓✓	✓✓✓	✓	✓
Laser radar	✓	✓	✓✓	✓	✓✓✓	✓
Laser tracker		✓	✓✓✓	✓	✓✓	
Laser tracker with tethered scanner	✓	✓	✓✓	✓✓	✓✓✓	

ever your application may be, it is the mission of this book to familiarize the reader with the basic principles of the equipment, software, and post-processing in the world of 3D laser scanning technology.

1.7 MILESTONES OF METROLOGY

The most important milestones of recent metrology history.[2]

Date	Event
1670	Proposition for a new length unit based on the terrestrial meridian 1799 Creation of the decimal metric system. Two platinum standards, representing the metre and the kilogram manufactured

---

[2] *Fanton: A brief history of metrology: past, present, and future,* 2019

Date	Event
1832	Austrian mathematician Gauss strongly promotes the application of the metric system, together with the second defined in astronomy, as a coherent system of units for the physical sciences; first measurements of the Earth's magnetic field take place
1860	Maxwell and Thomson formulate the requirement for a *coherent system of units* with *base* units and *derived* units
1880	Approval by IEC of a mutually coherent set of *practical units*. Among them were the *Ohm* for electrical resistance, the *Volt* for electromotive force, and the *Ampere* for electric current
1875	Signing of the *Metre* Convent ion, which created the BIPM, established the CGP M and the CIPM, and adopted the MKS system
1889	The first conference of CPGM takes place
1901	The so-called rationalized proposal of Giorgi, for a single coherent four-dimensional system, by adding to the three base units a fourth unit of an electrical nature such as the Ampere or the Ohm, and rewriting the equations occurring in electromagnetism
1939	Adoption of a four-dimensional system based on the metre, kilogram, second, and Ampere, and the MKSA system, a proposal approved by the CIPM in 1946
1954	Introduction of the Ampere, the Kelvin, and the Candela as base units, respectively, for electric current, thermodynamic temperature, and luminous intensity
1960	The name *International System of Units*, with the abbreviation SI, is given to the system
1971	Introduction of the last SI base unit: the mole, as the base unit for amount of substance, bringing the total number of base units to seven
1999	Signature of the CIPM-MRA (Mutual Recognition Agreement), for international recognition of national measurement standards
2018	New definition adopted concerning four base units on 7 (Meter, kilogram, second, Ampere, Kelvin, Candela, and mol)

REFERENCES

Brookes, J. (2015). The measure of all things: a brief history of metrology. AZoM. 07 September 2022. https://www.azom.com/article.aspx?ArticleID=12035. (accessed February 16, 2022).

Bucher (2004). *The Metrology Handbook*. ASQ Quality Press.

Fanton (2019). A brief history of metrology: past, present, and future. *Int. J. Metrol. Qual. Eng.* 10 (5).

Hocken & Pereira (2017). *Coordinate Measuring Machines and Systems.* 2nd Edition. CRC Press.

Sawage, M. (2013). *The Origin and Evolution of Calipers.* Mitutoyo.

Chapter 2
3D Scanning Basics

2.1 BASICS OF LASER LIGHT TECHNOLOGY

Today we find lasers being used in a variety of ways in industry. From light shows and CDs to cutting metal, and cauterizing wounds, lasers have earned their way to becoming an invaluable tool in today's world.

3D laser scanners have found their way to Hollywood as advanced technology with a "WOW" factor sought after by TV and movie producers. It gives us a futuristic glimpse of what science and crime fighting will be like. Who knew metrology could be so fun?

Lasers has been around for several years now and continues to improve. Various materials and designs now have lasers small enough to be soldered to a circuit board with their light beam channeled through a maze of fiber optics. But before we go any further, let's discuss these futuristic devices and learn a little bit about lasers, safety concerns and how they work for metrology applications.

2.2 LASER SAFETY

Lasers used to measure are generally considered safe though some are safer than others by design. Lasers used in high precision metrology are tuned to measure smaller objects with more detail. This is done using a more focused laser beam programmed to scan smaller details at high frequencies. Thus, the target gets more radiation from

3D Scanning for Advanced Manufacturing, Design, and Construction, First Edition.
Gary C. Confalone, John Smits, and Thomas Kinnare
© 2023 John Wiley & Sons, Inc. Published 2023 by John Wiley & Sons, Inc.

the laser than a broad swept target typically observed with overhead LIDAR surveys. Both instruments operate using visible and IR (infrared) wavelengths. The IR beam is normally the data or "information" beam and the visible beam is used to indicate direction and target acquisition (and safety) (Figure 2.1).

Laser safety is dependent on a few variables; wavelength, intensity, time on target. Intensity for measurement survey instruments is generally very low when compared to night vision systems used by the military or cutting and marking lasers used in manufacturing. The safety impact of these instruments mostly lies within the wavelength of the laser. IR light is a wavelength that is undetected by the human eye making it unsafe because you can't control the time of exposure. Think of the sun; the sun emits many wavelengths of light, but it is the visible wavelength that causes you pain and makes it impossible to do damage unless you deliberately stare into it. The sun has a built-in safety mechanism. For the same reason, industrial lasers will carry a visible beam so that the operator is aware of the exposure to the nonvisible IR beam and will be forced to blink or look away. It's the same principle as putting a scent in propane so that you can detect the otherwise scentless gas. So, if a Laser pointer was directed at your eye and you were immobilized you would probably be forced to blink because of the visible beam, but if you did not blink you could cause damage to your retina. (Note, laser pointers are actually more powerful than most lasers used in measurement devices.) Laser scanners used for large area data collection (architecture, historic preservation, land surveying, etc.) may come in contact with your eyes for a brief moment, but even with the visible carrier beam, the amount of time these broad sweeps would be in your line of sight would not be enough to cause any damage. Lasers are classified based on their safety factor. Below is a table illustrating the classifications and the precautions that should be followed when using lasers. Do note however, most lasers used for metrology are Class 2 or 3 which are considered safe in most circumstances. Always check the classification of the laser you are using – typically listed on outside tag or label of device.

Class 1 Lasers
Safe under all conditions, and Class 1M is safe under all conditions except when passed through the lens found in telescopes or microscopes.

Figure 2.1 Laser radiation warning sign

Class 2 Lasers

A Class 2 laser is a safe laser as long as the eye is not exposed to the laser for longer than .25 second. Often the eye will blink prior to the .25 second limit as a natural defense. But, as a safety precaution Class 2 lasers must include a label warning communicating something along the lines of "Do not stare into beam." Similar to Class 1, Class 2 also has a subclass called Class 2M which labels a laser safe with the .25 second limit excepted when passed through a magnifying lens.

Class 3R/3A Lasers

For a laser to be considered a Class 3R laser it must be under 5 mW. These lasers are considered safe if beam viewing is restricted. Making eye contact with the beam puts users at a low risk of injury. Class 3 lasers must include a label warning communicating something along the lines of "Avoid direct eye exposure."

Class 3B Lasers

When in the presence of a Class 3B laser the potential for serious injury reaches a medium/high level. It's very common for protective glasses to be worn when using Class 3B lasers and are required to include a warning label reading "Avoid exposure to beam." You may be surprised to learn that, while they are the second most powerful, Class 3B lasers are used to read CDs and DVDs.

Class 4 Lasers

Class 4 lasers are the most powerful lasers with the highest potential to cause severe harm to humans. Unlike the other laser classes which only deem a laser beam dangerous if *directly* viewed, a Class 4 laser can cause severe damage even if *indirectly* viewed. They have the ability to burn skin if exposed and present a fire risk. Laser safety is of utmost importance when handling Class 4 lasers.

Table of Laser Safety Classifications

Class	Description
Class 1	Safe under reasonable operation.
Class 1 M	Generally Safe. Some precaution is necessary.
Class 2	<1mW average power, visible light low power. Limited risk due to blink response.
Class 2M	UV or IR light at low average power. Generally safe LED systems.
Class 3 R(A)	Safe for viewing with the unaided eye.
Class 3a	1–5 mW of average power. Safe if handled carefully. Avoid staring into the beam.
Class 3b	5–500 mW of average power. Viewing beam is hazardous. Diffuse reflections are safe.
Class 4	>500 mW average power. Hazardous to eyes under all conditions.

2.3 SO WHAT EXACTLY IS A LASER?

Light **A**mplification by **S**timulated **E**mission of **R**adiation

Laser light is an electromagnetic energy much like a radiowave or a microwave. Below is a diagram (Figure 2.2) illustrating where light waves are located on the Electro Magnetic (EM) Spectrum. Most light sources generate white light which is composed of multiple wavelengths of energy. Recall using a prism in high school physics class to separate and disperse the seven colors of the spectrum from a white light source: red, orange, yellow, green, blue, indigo, violet. Each one of these color bands represented a specific wavelength of light. A laser is typically defined as a coherent beam of light based on a stimulated emission of electromagnetic radiation. This translates to the fact that a laser is a single wavelength of light just like one of the beams from the prism experiment. This beam is collimated using an array of lenses to create a long narrow beam that will travel great distances with minimal dispersion. By collimating the beam you preserve the energy at greater distances from the source.

The varying wavelengths (colors) of laser light are dependent on the medium used to generate the laser beam. For measurement purposes where a precise collimated beam (an aligned beam of parallel light particles) is desired, the optimal wavelength resides in the infrared wavelength region of the EM Spectrum. Generally, beams that are wavelengths of 560–632 nanometers are used for these applications. These small wavelengths have the ability to pick up surface details without penetrating the object.

Thus, if you were trying to transmit a signal through clouds or similar objects, you would not use a laser beam. This is where radio- or microwaves have an advantage for long-range transmission and detection. At these wavelengths the signal beam has the ability to penetrate clouds and other objects. On the other hand, X-ray beams are a form of electromagnetic radiation with a very short wavelength, much shorter than visible light. Unlike visible light, which is reflected from most objects, X-rays can penetrate through objects.

Multiple laser technologies are often used in many pieces of machinery and equipment to assist in their operation. Autonomous vehicles will employ several

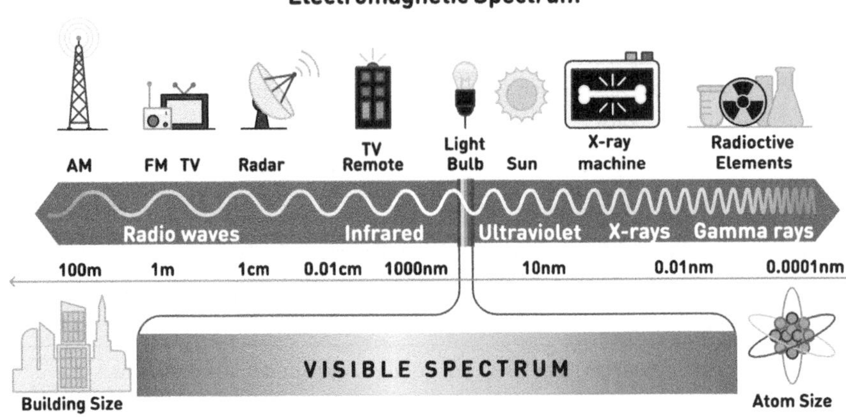

Figure 2.2 Electromagnetic spectrum

technologies in order to "read" the road and determine the threat. The limiting parameter of LIDAR is the line-of-sight issue. The wavelengths used for LIDAR will pick up detail but will not penetrate objects. This is where RADAR has its advantages. Different frequencies of electromagnetic waves are capable of reflecting or penetrating various objects. [Think of clouds captured by weather RADAR.] Using a combination of LIDAR, RADAR and Vision Systems (cameras) is necessary to direct these vehicles through real-world scenarios. The more frequencies you have, the more you can "see." Of course the detection is just a part of the puzzle. It is the data processing that really is the brains of the operation, sorting and making sense of the environmental data gained.

2.4 LASER TYPES

Next let's look at the different types of lasers and their basic principles of operation. There are essentially five different types of lasers commonly used in manufacturing today:

- Gas Lasers
- Liquid Lasers
- Semiconductor Lasers (Laser Diode)
- Solid-State Lasers
- Fiber Lasers

The Gas Laser, Semiconductor Laser (Laser Diodes), and Solid-State Laser are most popular within the metrology industry. A brief introduction to all five of these lasers and their operative qualities follows:

Gas Lasers

The two types of lasers primarily used for dimensional measurements are gas lasers and semiconductor lasers (laser diodes). A gas laser basically consists of a source, a lasing medium, and two mirrors (Figure 2.3). The source will stimulate the medium with an electric charge or photon energy. The medium is a gas or substrate that has the lasing capability when stimulated by the source. If the source is in a tube (picture a fluorescent light or neon sign), the mirrors would be located at each end. One mirror placed at one end of the tube has 100% reflectivity and the mirror at the other end of the tube has 95% reflectivity.

The most common gas lasers used in metrology utilize a medium of Helium Neon (HeNe 632.8 nm). The source excites the particles in the medium causing them to emit photons. These photons excite other particles causing them to emit more photons. The mirrors reflect the photons back and forth creating even more photons. This produces what is called a "photon avalanche."

Because one end of the tube has a mirror with 95% reflectivity, all of these photons will essentially bounce back and forth 95 times until the beam finally passes through

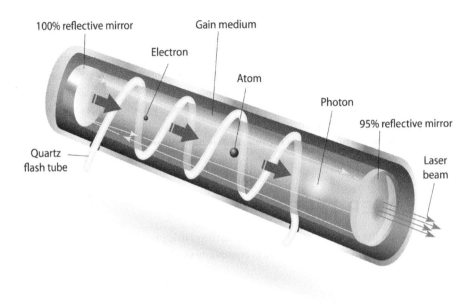

Figure 2.3 Cross-section view of gas laser

the mirror producing a coherent beam of light called the laser beam. Like all lasers, this beam is not perfectly collimated, so optics are used to control the divergence of the beam for various applications and focal length requirements.

Liquid Lasers (Dye Lasers)

The medium of a liquid laser is made from a concentration of dye molecules (typically in powder form) which is dissolved into a solvent. The medium is stimulated by optically pumping using a laser diode or flashlamp. Because Liquid Lasers use various organic dye molecules, these lasers have the ability to operate over a wider gain spectrum ranging from ultraviolet, visible, to near infrared wavelengths (30–50 nm). This tuning capability makes these lasers very useful in spectroscopy, a science used to identify materials based on the interaction with various electromagnetic wavelengths.

Semiconductor Lasers (Laser Diodes)

A laser diode is a semiconductor laser that has displaced the gas laser in a lot of applications because of its size and robust design characteristics. Modern gains in technology have allowed the use of the laser diode in precision measurement tasks due to the improved wavefront properties which were previously only found in gas lasers.

A laser diode is similar in design to a simple diode with the exception of a substrate layer (intrinsic layer) between the two silicone layers labeled p-type and n-type in Figure 2.4. When an electric charge is placed across these layers the intrinsic layer in the middle

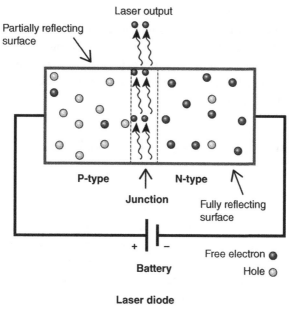

Figure 2.4 Cross-section view of semiconductor laser (physics-and-radio-electronics.com/ n.d.)

becomes excited when free electrons move across this layer from the n-type to the p-type producing charged particles and thus releasing photons in the process. In this process each photon will create two more photons and so forth. The surfaces on either side of the intrinsic layer are polished to a mirrorlike finish causing the particles to bounce around like the particles of a gas laser. Eventually the particles will create a beam departing from the substrate in the middle of the laser diode on the partially reflective side. Lenses and fiber optics are used to shape the laser and channel it as required.

Solid-State Lasers

Solid-state lasers operate using crystals mixed with rare earth element as a gain medium. This gain medium is stimulated by "pumping" it with an electric charge or a light source. Most often this pump light source is a laser diode. The first laser ever constructed was a solid-state laser which used a ruby as a medium. Today the most common solid-sate lasers used are the Nd:YAG laser (neodymium-doped yttrium aluminum garnet).

In metrology, solid-sate lasers are most often used for Airborne LIDAR applications. In this situation, the LIDAR system is mounted in a drone or aircraft where it is able to pulse a laser beam at a building or at the ground and determine its exact location by the time it takes for the reflected beam to return. This is accomplished using the mathematical equation:

Distance = ½ (Speed of light × time of flight)

Fiber Lasers

A fiber laser is a type of solid-state laser using a doped fiber optic as a medium. This is done using a silica fiber mixed with a rare-earth element. Though similar to the solid-state laser, the unique light-guiding properties of fiber lasers allow them to be used in a variety of applications including optical fiber communications, laser surgery, LIDAR and range-finding, and laser pumping.

2.5 LIDAR VS LASER-RADAR

In the world of 3D scanning, confusion between the two is common. Their names are nearly synonymous, and the terms are often used interchangeably. The acronyms are RADAR, which stands for RAdio Detection And Ranging; and LIDAR which stands for Light Detection And Ranging. The major difference being the wavelength of the signal and the divergence of the signal beam. LIDAR is typically a collimated light beam with minimal divergence over long distances from the transmitter, where RADAR is a cone-shaped signal fanning out from the source. Both calculate distance by comparing the time it takes for the outgoing wave or pulse to return to the source. LIDAR uses light wave frequencies which have a shorter wavelength enhancing the capability of collecting data with high precision. RADAR uses longer microwave frequencies which have lower resolution but the ability to collect signals with reduced impact from environmental obstructions. RADAR and LIDAR signals both travel at the speed of light.

LIDAR is used as a generic term for most light-based metrology technologies. Typical devices use laser beams, structured light mesh, even pulsed white light, based upon the system's design and application. All are used in radar-like fashion to measure the position of target points. They typically measure in three dimensions by monitoring a vertical angle, a horizontal angle, and the distance or range, from the intersection of the vertical and horizontal axes. Similar to RADAR, some LIDAR systems measure in only two dimensions by monitoring a single angle and the range.

Knowing the similarities between RADAR and LIDAR let's now take a look at LIDAR versus Laser-Radar. By definition, LIDAR and Laser-Radar refer to the same principle methodologies of measuring an objects position. In the recent past, the term Laser-Radar was adopted to define measurement systems designed to collect data with extremely high precision. This is done by using a narrowly focused light source of a specific wavelength or a combination of wavelengths. The low divergent beam is capable of focusing on small details with high return resolution.

For comparative purposes, let's take a closer look at two laser-based scanning units.

2.6 LASER-RADAR

Figure 2.5 displays a "Laser-Radar" system in which the laser beam is guided by the unit rotating on both horizontal and vertical axes. This unit sends a continuous frequency

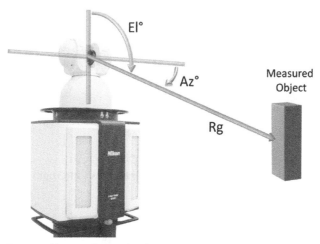

Figure 2.5 Focused beam Laser-Radar system
Source: Image courtesy of Nikon Metrology Inc.

modulated, focused laser beam to an object and analyses the return signal to determine range very accurately from the relative phase shift. Unlike laser tracking and other surveying instruments, it does not need to use a retroreflector. Its signal is the reflected light from the actual objects surface. It is engineered to provide precise, industrial measurements with tolerances of thousandths or even tenths of thousandths of an inch. The units have an effective range of measurement up to 150 feet in radius, although in practical use the range is usually shorter. Rather than taking large sweeps of an object, the Laser-Radar system takes smaller scans of areas where high accuracy and detail is the priority. Because of this, speed of data collection is sacrificed for resolution in comparison to LIDAR systems. Automated inspection of difficult to access areas, or time sensitive measurements in manufacturing production lines have proven to be the most useful applications of these systems.

2.6.1 LIDAR

Figure 2.6 illustrates a laser scanner, a phase-based long-range laser scanner or "LIDAR" system. This is a typical 3D scanning unit used for large-scale data collecting at high speed. The laser emitter is fixed horizontally within the left half of the case and is pointed at the center of a rotating angled mirror that pulses a continuous series of laser beams into a plane perpendicular to the common axis of the laser and mirror. The entire case rotates on the base to provide measurements within a 360-degree horizontal arc. The resulting pattern of laser data produces what is referred to as a point cloud, a series of data points with x,y,z information, creating a digital 3D environment of any object or landscape which is scanned. The LIDAR system is capable of collecting large amounts of data in a very short amount of time.

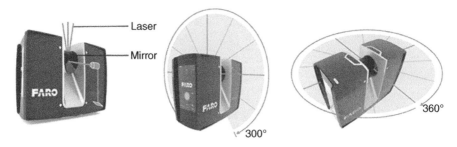

Figure 2.6 Phase-based laser scanner or LIDAR system
Source: Image courtesy of FARO Technologies, Inc. www.FARO.com.

These two devices exemplify a key difference between LIDAR and Laser-Radar: a LIDAR system is usually constantly in motion while gathering measurements. LIDAR continuously sweeps a laser beam over a large area to gather millions of points while the head is moving at a rapid speed. Laser-Radar is more of a point, stabilize, and measure device. It can be used to scan an area, but the points are carefully chosen, acquired slowly and more accurately.

The cardinal distinction in these two systems is mostly acknowledged through applications. The term LIDAR is used frequently in surveying and mapping. Attach a laser scanning LIDAR system to an aircraft or automobile, combine it with GPS or another orientation system, and large swathes of terrain or roadside can be mapped. Other LIDAR examples are drone-mounted scanners or self-driving car guidance systems. The required accuracy for this kind of work can range from a tenth of an inch to a foot; ranges can be thousands of feet provided there is unobstructed line-of-sight to the target. Portable "long-range" 3D laser scanners that measure rooms, buildings, aircraft, ships, or similarly sized objects to accuracies of 0.030–0.120″ are also in the LIDAR category. This is the kind of LIDAR technology often used in construction or engineering.

Laser-Radar systems have found their value in measuring finely detailed objects. Their tightly focused beam is capable of collecting precise data points with high levels of resolution on a specific detail of an object. By minimizing the contact area, these systems will reduce the chance of corners and edges adversely affecting the return signal, resulting in precise dimensional data. Reaching targets that are difficult to access with handheld scanning devices has proven to be the most useful application of these systems. With accuracies in the realm of portable coordinate measuring machine (PCMM) devices the Laser-Radar data is perfect for reverse engineering, inspection, and Part-to-CAD comparison. Small details such as hole locations and diameters are easier to obtain with the precise focused beam of Laser-Radar units. These systems have found a niche in the aerospace and automotive industry where precision data is easily obtained with minimal intrusion.

2.7   BASICS OF 3D SCANNING

Throughout the years, laser-scanning technology has evolved and encapsulated a variety of optical-physics techniques for capturing data. To understand the physical components required to gather data in 3D space, let us start with the basic principles and work our way to the most advanced systems known today. Simple yet sophisticated, laser-based metrology devices have found their way into our lives in the numerous practical applications. Laser rangefinders are a prime example, their uses are wide and varying. They are capable of recording accurate distances for a variety of applications. Architects, builders, and interior design professionals use them for dimensioning rooms and buildings. Machines and equipment utilize a rangefinder to assist its operation and spatial detection. The laser beam is the foundation of many more sophisticated measurement devices, so it is suitable that we begin by explaining the basic principle of operation of these devices.

2.8   HOW IT WORKS

These time-of-flight devices operate simply by measuring the time it takes for a pulse of light to reflect off a target and return to the transmitter. The transmitter in this case being a laser distance meter or rangefinder.

Lasers are used for many reasons but primarily because they are essentially a single frequency, nondiverging beam of light. What this means is that the light beam is able to maintain its intensity while traveling through the atmosphere with minimal dispersion. And, unlike white light, a laser is able to maintain its intensity when reflecting off a target. This is particularly important when measuring distance.

Knowing that laser light will travel through the air at a near constant speed, we are able to calculate the distance from the source to the target. The distance is calculated using the relation distance = (time) × (velocity)

$$Distance = (c \times t) / 2,$$

where "c" is the speed of light and "t" is the time it takes for the pulse to make the round trip from the source to the target, and back.

It is easy to understand the basic principles although the electronics of the device are somewhat intricate. These devices by nature are very inexpensive and not typically used for high accuracy metrology.

2.9   LIMITATIONS

Because light travels at such a high speed, it is difficult to collect measurements when very close to the source emitter. Thus, the systems are generally applied to measurements in feet rather than inches.

The wavelength of the laser and the divergence or focus of the beam will dictate the ability for the laser to measure out to far distances. It is very difficult and expensive to maintain intensity of a nondivergent, focused beam when measuring long distance, but it is not impossible. For example, the distance to the moon has been measured by astronomers all over the world within accuracy of a few millimeters, using this time-of-flight method by aiming a focused laser beam at a retro-reflective panel that was placed on the moon's surface by the Apollo 11 mission in 1969 (Figure 2.7).

Distance meters are also subject to background noise in the form of light from another source reflecting back at the source detector. This will cause inaccurate or false readings from the device. To minimize this effect, bandwidth filters and split beam frequencies can be applied to filter out the interference from unwanted background light.

2.10 ACCURACY

Like all light traveling through the atmosphere, it is subject to distortion which will make it difficult or impossible to obtain an accurate measurement. Temperature gradients are the biggest cause of error when using laser distance meters. These effects are the same as one would see with white light distortion on a dessert highway. This refraction of light will not allow the laser pulse to travel in a straight line from the source to the target causing a distance error. Also, the target surface finish or material may have properties that will absorb the light reducing the return signal strength. Black or carbon fiber finish materials can often present this challenge. Adversely,

Figure 2.7 This Retroreflector was left on the Moon by astronauts on the Apollo 11 mission. Astronomers all over the world have reflected laser light off the reflectors to precisely measure the Earth–Moon distance
Source: NASA.

highly reflective surface can scatter or diffuse the signal which will also make it difficult for the light pulse to be reflected back to the source. In these instances, a phase shift distance meter should be used.

2.11   MODERN METROLOGY

Large volume metrology took the next big step when lasers were used in optical instrumentation to measure distances. Using two positions to triangulate on a target was not always required with the integration of range detecting laser technology to the total stations. Moreover, the laser tracker revolutionized portable 3D metrology using interferometric range capturing technology as discussed in Chapter 3.

REFERENCES

Beheim, G. and Fritsch, K. (1986). Range finding using frequency-modulated laser diode. *Appl. Opt.* 25: 1439–1442.

Bosch, T. et al. (1992). The physical principles of wavelength-shift interferometric laser rangefinders. *J. Opt.* 23: 117.

Confalone, Smits, Kinnare (2020). What's the difference between Laser Radar and LIDAR technology, Quality Magazine, May 2020.

Hayt, W.H. and Buck, J.A. (2012). Engineering electromagnetics, McGraw-Hill Education.

Lawrence Berkeley National Laboratory, U.S. Department of Energy National Laboratory. (2018). Laser classification explanation. https://ehs.lbl.gov/resource/documents/radiation-protection/laser-safety/laser-classification-explanation. (accessed March 8, 2022).

Chapter 3

Scanning Equipment

3.1 INTRODUCTION

Laser scanners use a variety of different technologies or combinations of technologies to detect surface points and record their characteristics. There are multiple varieties and brands of equipment that will produce 3D datasets. Selection of which scanner and methodology to be used is based upon the type of object to be scanned, level of precision required, and the final deliverable needed. Other factors such as the scanner's cost, size, and portability influence the choice of equipment.

Laser scanners that produce point cloud datasets can be categorized by their operating technology into four main types: triangulating laser scanners, LIDAR systems, structured light systems, and photogrammetry systems. Other methods of scanner classification is sorting by common technical and physical attributes. Examples of this are devices sorted by range: short, medium, or long; or by type: handheld units, tripod mounted, or desktop units. To select a scanning technology or specific scanner suitable for a project, it is important to understand how the various technologies work and interact with the objects to be measured.

3.2 TRIANGULATING LASER SCANNERS

Triangulating laser scanners is a generic title given to any scanning device that measures data points through triangulation. This is a term that can also be applied

3D Scanning for Advanced Manufacturing, Design, and Construction, First Edition.
Gary C. Confalone, John Smits, and Thomas Kinnare
© 2023 John Wiley & Sons, Inc. Published 2023 by John Wiley & Sons, Inc.

to some structured lights systems and photogrammetry. The operating basis of any triangulating laser scanner can be pictured schematically as seen in Figure 3.1. Laser light is emitted from the source, bounces off the measured surface, and returns to be captured by the camera or detector. The distance between the object being scanned and the scanner is calculated using trigonometric triangulation. The laser and the return sensor, or detector, are set at fixed distance apart, and the angle between the two of them is also fixed. If the surface is close to the detector, the returned laser light will register higher on the detector. If the surface is farther from the detector the light will register lower. By registering this return point's position on the detector, the scanner calculates the angle of the return, providing the information required by the algorithms to establish the location of the surface point relative to others.

The image sensors in the scanner typically have a small dynamic range and are best used within close proximity to the object, usually within five meters. In actual practice there may be several lenses, mirrors, and filters in the laser's path to aim, spread, focus, orient, or otherwise manipulate the laser light as needed. Left to itself, a laser will project a dot onto the surface. Most scanners use lenses or an oscillating mirror to turn that dot into a line, which enable better measurement results. Optical adjustments make the laser line as wide as possible. In general, wider is better; a wider laser line means fewer scan passes which means less time measuring. However, wider laser lines generally correlate to larger, heavier instruments which are unwieldy to work with, especially when working with small parts or objects. The return recorded on the image sensors is ultimately channeled through a processing chip in the scanner which converts the raw detector data into coordinates to be fed to the data-acquisition system.

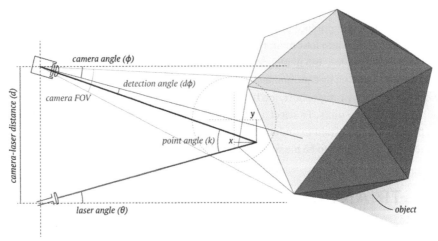

Figure 3.1 Typical operating basis of a triangulating laser
Source: Image courtesy of Artec3D, www.artec3d.com.

3.3 DATA ACQUISITION VARIABLES

The data return from the laser will be affected not only by the strength and focus of the laser, but by the surface which the laser strikes. There are primarily two main types of surfaces encountered, specular (mirror like) or diffuse (Figure 3.2):

- *Specular surfaces.* When light strikes a mirrorlike surface almost all of it bounces off in a concentrated line. A very small amount of the light is scattered in other directions. If the laser beam is perpendicular to the surface, then the light is reflected straight back to its source. This is referred to as the "normal." If the light strikes the surface at some other angle from the normal of the surface, that is referred to as the angle of incidence. Depending upon the degree of the angle of incidence, a portion of the light is reflected away from the surface at the same angle, mirrored around the normal.

- *Diffuse reflecting surfaces.* Diffuse reflecting surfaces scatter incident light, the light that falls on a surface. An ideal diffuse surface breaks up any incident laser light and sends a portion of it in every direction. Such surfaces often have a flat or matte, as opposed to glossy or semiglossy, finish. Matte surfaces are often rough when viewed on a microscopic scale. The roughness of the surface causes the incident light to strike the surface at many angles, which in turn causes the light to scatter.

3.3.1 Real Surfaces

Real surfaces exhibit a combination of specular and diffuse reflective characteristics. A portion of any incident light hitting the surface is reflected in a mirrorlike manner, and the remainder scatters. The physical shape of the object's surface also affects the return of light to the scanner. Typical surface characteristics and issues affecting return of laser data are:

- *Angle of incidence.* Ideally it is best to measure a tiny point or very thin line of laser light on the object's surface. The smallest possible spot or line occurs when the laser is being projected at 90 degrees to the surface, when the angle of incidence is zero. As the angle of incidence increases, a circular spot turns into an

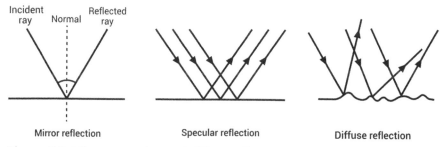

Figure 3.2 Mirror, specular, and diffuse reflection

oval; a line of laser light thickens. If we measure a large spot or thick line then a number of possible errors can creep into our measurement. Some are small; some can be significant. The worst probably occurs near edges.

- *External corners* cause a portion of the laser light to split. It does not fully hit the object, so a portion of the laser beam is not reflected back to the detector. Instead the detector registers the portion of the beam that is reflected and measures the point at a false position farther from the scanner than the actual part surface.

- *Interior corners* can cause a laser light to reflect from the objects connected sides. This raises the apparent center of the laser line as seen by the detector off the surface, and a point is recorded closer to the scanner than the actual surface.

- *Line of sight.* If anything interrupts or breaks the laser beam, it will not record and measure any point data. That applies both to the line of laser light projected onto the object's surface, and the reflection from the surface back to the detector. Normally, line of sight issues are clear to the operator. However in some instances operators don't recognize how a step, hole, or crevice in the scanned object may be blocking the light's path back to the detector. Problems involving line of sight clarity will result in voids within the overall object's scan data.

- *Scanner proximity to the scanned object.* A scanner is designed to have an ideal distance between the scan head and the object to be measured. Typically, at this distance light reflects back to the center of the detector. If you move the scanner closer to the measured object, the reflection moves up the detector. If you move the scanner too close, the reflection falls off the detector and you stop collecting data. Likewise, if the scanner is farther away the reflection moves down the detector until it falls off.

  The strength of the laser beam is also a factor in recording measurements. Too far away and the laser will not send back a light beam able to be recorded. Too close and the resulting return could overload the data sensor. Every scanner has an ideal range (distance from the scanner), and a span (range) around the ideal distance in which the instrument will collect data.

The primary issue affecting data return is the angle of incidence. The resolution of the scanner changes and is reliant on the detection angle, which is the angle between the local normal of the measured surface and the light reflecting from the surface (reflected ray) back to the detector. Refer to the diagram of the "mirror reflection" in Figure 3.2. These changes in angle and the resulting effect on the data can be mathematically quantified. Table 3.1 charts the laser light returning to the detector for a 0.001 inch deviation in surface position. This deviation will change $.001 \times (\sin \phi)$, where ϕ is the angle between the local surface normal and the reflected ray.

The sensitivity of the detector increases as the sine of the angle increases. Thus, $\sin \phi$ would equal 1 if the detector was pointed at 90° to the measured surface. This would not be very practical however because any imperfections that raised the surface between the laser line and the detector would block line of sight. Pockets or concavities in the object will therefore be difficult to measure.

Table 3.1 Camera sensitivity vs angle from normal

Camera Angle from Surface Normal	Sine	Change at Detector from .001 Change in Surface
10	0.1736	.00017
20	0.3420	.00034
30	0.5000	.00050
45	0.7071	.00071
60	0.8660	.00087
75	0.9659	.00097
90	1.0000	.00100

The factors affecting operation of triangulating scanners can be applied to all scanning groups. Angle of incidence, proximity of the scanner, and surface type will influence data quality in all types of scanners.

3.4 SCANNING EQUIPMENT CATEGORIES

The following is an introduction to some of the most commonly used scanning equipment, their primary functions, and applications.

3.4.1 Coordinate Measuring Machine (CMM)

The coordinate measuring machine (CMM) has long been the standard benchmark for industrial metrology. A highly accurate device, the uncertainty associated with a single point measurement may be on the order of .0002″. Its measurements became the prevailing means of checking for quality assurance of manufactured parts. The CMM measures the physical geometry of an object. A probe attached to the end of the CMM machine arm is used to touch the object, recording a single xyz data point. Multiple points placed upon the object create a dimensional dataset that is used to verify its adherence to the intended design configuration (Figure 3.3).

A CMM is mounted on a rigid heavy tabletop, usually made of granite, steel, or other dense material. This benchtop is where the objects to be measured are placed. It provides a secure base resistant to movement or vibrations, allowing for dimensions to be taken unaffected by external forces. Mounted above this plate is a movable gantry; attached to that gantry is a vertical column upon which a measuring probe is mounted (Figure 3.4). The gantry, along with its descending vertical arm, is able to move the probe along

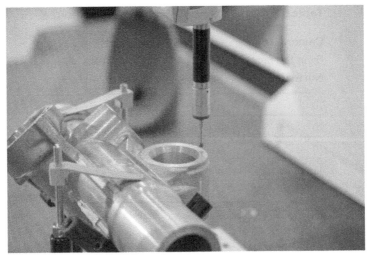

Figure 3.3 Probe tip of a CMM measuring a machined part

Figure 3.4 Bridge CMM unit with part ready for inspection

a three-axis path. Each time the probe contacts the object an electrical pulse is sent to a computer, recording an xyz point on the surface of the object. The computer is part of the CMM's control system; it contains software to run the machine, sets a coordinate system to determine the exact relative location in space of all touch probe points, and produces a drawing or data file of the points recorded. Depending upon the size of the object, multiple touch points are taken. The ensuing file generates a point cloud of the object's surface configuration. This file can be analyzed against the object's 2D or 3D CAD design files to ensure the part has been produced as intended.

When producing tight tolerance machined parts, it is important that all pieces fit together properly and meet the intended design. The CMM's inspection process will not only ensure parts meet specification, but can identify potential flaws in the part, allowing either a redesign or change in the production process. CMM machines vary in size and operational abilities. They are typically one of four types: bridge, gantry, cantilever, or horizontal arm. Each one is designed to serve the specific logistical measurement difficulties caused by an industry's unique production pieces. All operate in a 3D coordinate system with three orthogonal axes, x, y, and z.

3.4.1.1   Bridge-type CMM

Bridge-style CMMs are the most common and are split into two types: moveable tables and movable bridges (Figure 3.5). These units provide a heavy stable base to hold the measuring device. These machines can be fixed in place, and their robust design allows them to withstand the rigors of a production floor environment. They are not as fragile as a laboratory machine. This style of CMM is best suited to measure small- to medium-sized parts which can easily be set upon the table where they are studied.

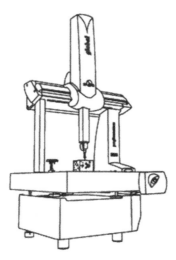

Figure 3.5 Bridge CMM
Source: Courtesy of Hexagon Manufacturing Intelligence, Inc., www.hexagon.com.

Figure 3.6 Gantry CMM

Source: Courtesy of Hexagon Manufacturing Intelligence, Inc., www.hexagon.com.

3.4.1.2 Gantry-type CMM

Gantry-style CMMs are designed for quality control measurements on large production items or heavy machine parts, such as an automobile or large aircraft engine (Figure 3.6). These machines use the floor as a base, eliminating the need to have a tabletop. Their specialized design allows large parts to be tested on the production line or brought to a nearby testing area, eliminating the need to lift or rotate heavy objects. They are the most expensive of the CMM types due to their size and installation costs.

3.4.1.3 Cantilever-type CMM

Cantilever CMMs are usually smaller units and used primarily for measuring of small parts. They have only one point of support, which gives them greater mobility in moving about an object. While the two aforementioned CMM types can access an object from two sides, the cantilever units add a third face of access. This is especially helpful when working with small intricate machined parts.

3.4.1.4 Horizontal Arm

The Horizontal Arm CMM is designed to serve the unique constraints of measuring long slender parts that exceed the range of a standard CMM (Figure 3.7). It uses a horizontally mounted probe that can travel along a longer surface distance for touch probing. This stands in contrast to the typical CMM which uses a vertically mounted probe and is typically designed to accept larger parts that would not fit on a Bridge CMM.

While CMM units differ in size, accuracy, and design, they all share common advantages. Advantages to using the CMM machine for quality control include:

- CMM machines are reliable; precision instruments calibrated and certified to typically be accurate within 0.0001.

- The machine can be hand operated or have a set pattern of dimensional point locations programmed into the machine. This enables applications to use for one-off prototype models as well as large-scale repetitive quality control on production items.

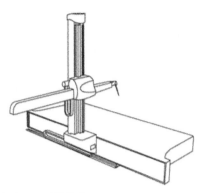

Figure 3.7 Horizontal arm CMM
Source: Courtesy of Hexagon Manufacturing Intelligence, Inc., www.hexagon.com.

- The machine can provide dimensional analysis where needed, whether it's on a portion of the piece to be inspected or the entire object.

- The machine's probe can be fitted with different stylus tips depending on the part to be measured. Various stylus tip designs provide the ability to measure diverse geometric features and surfaces.

- CMM units take up limited space and can be conveniently located on the production floor. This avoids parts having to be moved off the floor to a lab.

As computer software and hardware expand its ability to handle greater datasets and complex modeling, industrial part production has grown increasingly complicated. In particular the use of parts or objects with multiple curved surfaces. This has given rise to the need for thousands of points to be taken on a surface in order to create a dataset useful for confirming design adherence. Simpler parts that once took a day's work with a probe are now much more complicated and require weeks of effort to obtain the data needed.

The use of new plastic materials poses logistical issues when using a contact probe CMM unit. Softer, flexible parts are difficult to measure with a touch probe. When measuring rubbers or elastomers the slightest indentation to the objects surface from the probe tip will cause a deviation from its true surface dimension. Other parts or objects can be made from materials with a delicate surface that a stylus tip will scratch or damage.

To respond to these issues of complexity and material, CMM machines began to be retrofitted to enable use of a 3D laser scanning probe. Mounting a laser scanner onto the CMM unit gave the benefits of typical arm and handheld scanning units to easily capture complex organic surfaces quickly and efficiently. It also gave the added benefit of a highly accurate positioning system, and the ability of the scanner to be operated by machine programmed pathways, thus eliminating the use and inconsistences of a hand-operated scanner.

The accuracy of a laser scanner cannot match the precision and reliability of a touch probe. For high precision machine parts and other similar objects, the probe tip CMM will always be the best choice. However, when the components to be measured have complex geometry, multiple features and a variance tolerance within the .01″ range, laser probes become a viable and cost-effective choice. This is primarily due to their speed in documenting an object, the ability to capture over 15,000 points per second, and record whole geometries rather than a limited set of points.

Technologies continue to evolve allowing industrial manufacturing to produce parts and machinery of more complexity. The advanced use of automation in industrial work flows creates an ever-increasing need for rigorous quality control to maintain reliability in the production of repetitive objects. There will undoubtedly be similar advances in CMM equipment design, computing capacity, and speed to meet the challenges of modern advanced manufacturing.

3.4.2   Arm Scanners

These scanning devices are based on a similar technology to CMMs. The triangular scanner head or probe is mounted at the end of a multiaxis flexible jointed arm which allows movement around the object to be scanned. This allows for complete coverage of all surface areas of the object or part to be measured. The arm head, or probe, records a series of 2D profiles of the objects surface, while the base unit tracks the position and orientation of the scanner head. Together the data from both parts of the system results in a precise 3D data set of all points measured (Figure 3.8).

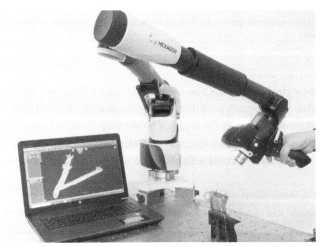

Figure 3.8 Typical articulated arm scanner with laser head mount
Source: Photo courtesy of Hexagon Manufacturing Intelligence, Inc. www. hexagon.com.

Articulated arm CMMs are smaller and are considerably less expensive than traditional CMMs. They provide many of the same benefits of a traditional CMM with a small loss of accuracy. Scanning arms are considered portable; they can be brought to a customer's site, installed on a sturdy work table surface or set in multiple locations around a part that is too large to fit within the reach of the arm's measurement range.

Many arms are manufactured with integrated mounts and electrical connections for scanners from the same manufacturer. The use of these single manufacturer systems is usually seamless, and the software can quickly and easily switch between hard-probing and scanning modes of measurement.

3.4.3   Handheld Scanners

Handheld 3D laser scanners have been developed for many applications across multiple industries. Most are small lightweight units that emit a beam or beams of light to record and enable measurement and documentation of the surface of an object. The data output from the scanners is in the form of point cloud data or a structured mesh. Scanners can be freestanding or connected to a computer laptop or tablet. The 3D modeling progress as the scanner moves across the surface of the object is displayed on a screen mounted on the scanner or on a connected laptop. By observing the modeling, the user sees which areas of the object are recorded and map the remaining scanning over the object. These units vary widely in cost, design, laser methodologies used, and accuracy levels. Prices of handheld scanners can range from a few thousand dollars to over $100,000. The more costly devices typically last longer with fewer problems and deliver scan data with better accuracy, density, and visual quality. However, a more expensive device may not be best for the intended use. When considering which scanner is best for a project, the following items should be considered:

1. What is going to be scanned? Will it be a small object with fine detail, or a larger object with minimal small surface variations? Will the data be used for quality control, replication, or reverse engineering? This will determine the accuracy level of the unit needed.

2. What kind of material is the object's surface made of? Dark or light, reflective or nonreflective? Some types of laser light used in a scanner often will not work well on dark or reflective surfaces.

3. Will the scanning be indoors or outdoors? Many units are best suited for low light levels and will have problems when scanning outside. In order for a scanner to be able to read the light return it may require a sunshade trap or umbrella. High temperatures will also affect scanner operation.

4. How easy is it to operate the scanner? Is it necessary to place markers on the object? Or around the object? Can the scanner keep track of its progress on the object with targets?

5. Will you work in a controlled environment or out in the field? How portable is the scanner?

6. How long can the scanner operate? Does it need to be connected to a laptop or tablet? What is the maximum time window for scanning before battery power runs out?

7. What kind of software is used by the scanner? Software for handheld scanners is typically proprietary by the device's manufacturer. Will it be updated regularly? What is the cost of maintaining current software?

Hand scanners usually capture data using one of two methods, structured light or time of flight.

Structured light scanning is a broad term for the process of recording the shape of an object by shining a precisely defined light pattern on the object and detecting the alterations in the pattern caused by the shape. In common usage, when structured light is mentioned, it is typically in reference to either accordion fringe interferometry (AFI) or what's commonly called blue or white light scanning. Accordion fringe interferometry projects two or more laser beams onto the surface to be scanned; the pattern is deformed by the surface, which is recorded by the scanner's detector.

Structured light devices, often referred to as a white light 3D scanner, are in some ways a combination of photogrammetry and triangulating scanning. In practical terms, compared to triangulating scanning, it's like a "wide view" scanner because measurements are taken over a wide area rather than along a single line of laser light. They work by sending a series of light patterns onto the object. These patterns can be white light, or other light spectrum colors dependent upon the scanner's design. The projected light pattern is produced by a combination of lenses, optical gratings, and shutters. A detector built into the scanner senses the shifts and change in the light's pattern as it crosses over the image scanned. A high speed camera, or pair of cameras, offset from the laser light source in the scanner take multiple frames per second to capture the deformities of the light grid caused by the object's surface geometry. The robust and exacting construction of the instrument holds the camera or cameras in constant position relative to the projector, and thus provides scale. Software in the scanner calculates the distance to the various points captured, creating a 3D image of the surface. As the scanner moves across the object, multiple frames are analyzed and stitched together to form a 3D digital image of the object. Monochrome cameras are standard in basic units; other units have color cameras that will generate a high resolution, visually correct representation of the object scanned.

In structured light scanners, the number of pixels, the camera resolution, is often an option. The customer may choose a higher or lower pixel camera. Higher pixel counts allow a scanner to distinguish finer detail and possibly increase accuracy. The lenses in front of the cameras can also be an option. Changing the focal length of the lenses will alter the effective size of the measurement volume and may affect accuracy. Changing lenses will also influence the optimum set up distance from the instrument to the object.

Handheld scanners that operate with laser beam mechanics similar to terrestrial scanners capture surfaces with a very narrow directed beam of laser light. These scanners can provide extremely precise measurements, within microns, way beyond the accuracy of a structured light scanner. They are used when accuracy of the data is of foremost importance. Often these highly specialized units are used in conjunction with a laser tracker or other device to assist in recording their tracking over an object. Their use of a red laser light can impede their use outdoors in bright sunlight areas.

Structured light systems are often used with a rotary stage. The instrument remains stationary while the stage rotates the object at a precise angle. The instrument, depending upon its design takes either incremental or continuous measurements. The process is repeated until the object has been rotated through 360 degrees. Circular reflective targets on the stage or object are often used to enable the instrument's software to accurately determine and align the object's angle of rotation and image capture.

When using a structured light system, like using a triangulating laser scanner, think about each line of structured light as a wedge-shaped probe whose narrow edge touches the part. It will be difficult to measure the inside of small holes or narrow slots because the adjacent part's surface will break the camera's line of sight into the hole or slot even if light from the projector is able to get inside. In some areas of the surface protrusions may cause shadows from the light pattern, inhibiting the detection of the surface anomalies. This can result in no point data collection in those areas, producing voids in the dataset.

Characteristics of the part surface, such as reflectivity, color, or translucence, may provide challenges for structured light similar to those faced by triangulating scanners. Corners and edges may provide multiple paths for light to reach the camera or allow partial reflections to deceive the camera. This resulting data noise can be resolved during the post-processing phase of the project.

The following are examples of handheld scanners and their specifications to provide an overview into the varieties of scanning devices available on the market. Keep in mind that most manufacturers produce multiple models of scan devices that will range in price and functionality.

3.4.3.1   Self-contained Scanner

Most handheld scanners rely on a connection to a computer to provide their data processing power. By building in the computer processing, it allows the unit to be free and untethered. This provides flexibility in moving around an object while scanning. A touch screen panel located on the unit provides the visualization need for the operator to follow scanning progress. One can even touch the screen and rotate the 3D image to ensure all areas have been adequately recorded.

The units are typically lightweight and designed to be held in one hand. They have an internal gyro and compass along with an accelerometer, a sensing device that measures movement forces to determine its positon in space. These attributes allow scanning without having to place targets or markers on or around the object. A color camera allows for high resolution imagery (Figure 3.9).

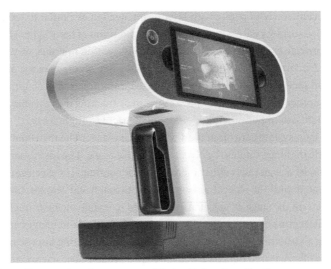

Figure 3.9 Self-contained scanner with built-in visualization screen
Source: Image courtesy of Artec3D. www.artec3d.com.

Typical specs:

Working distance range: 0.35–1.2 m
Volume capture zone: 160,000 cm³
3D resolution up to: 0.2 mm
3D point accuracy up to: 0.1 mm
Data acquisition speed up to: 35 million points per second
Color: 24 bit
Light source: VCSEL – vertical cavity surface emitting laser (a semiconductor-based laser diode)
Data output: multiple 3D mesh formats available (i.e. OBJ, STL, E57, PLY)

3.4.3.2   Tethered (Computer-connected) Units

Many tethered units attain the level of metrology grade precision instruments. They are extremely accurate and can be used in lieu of a CMM for certain types of quality control checks. The device must be connected to a laptop and power source, typically achieved by a long cord supplied by the manufacturer. The unit sends a grid of laser light across the object to be scanned. It is caught by the cameras in the handheld device which record surface deformities; the resulting data is then converted into a structured mesh which displays as a digital image on the laptop computer screen. It is highly accurate, capable of capturing minute surface variations. Due to its sensitivity levels and built-in tracking system, these units typically must be calibrated prior to using. This involves scanning a fixed target board, during which a series of commands in the software instructs the user through steps to verify accurate operation of the scanner (Figure 3.10).

Figure 3.10 Tethered unit with laser grid lines on propeller blade. Note small circular black and white dots to enable tracking of the scanner
Source: Image courtesy of Creaform/Ametek, www.creaform3d.com.

Typical specs:

Working distance range: 300 mm
Accuracy: 0.035
Scanning area: 310 × 350 mm
Measurement resolution up to: 0.025 mm
Mesh resolution up to: 0.100 mm
Data acquisition speed up to: 480,000 points per second
Color: 24 bit
Light source: Blue Laser
Data output: multiple 3D mesh formats available (i.e. OBJ, STL, E57, PLY)

3.4.3.3   Semi-Tethered Units

Some manufacturers have resolved the scanner's computer and battery connection issues by providing a small mobile computer unit as part of their handheld system. This unit is worn on the user's belt or a shoulder strap, along with a small cord that connects to the scanner, thus allowing full flexibility of movement to and around an object. It scans using a projected light grid from an array of LED strobes. Cameras record the warped grid points for processing, the center mounted camera is a high-resolution color camera that shoots the image frames. Some units have a tracking algorithm built into the software enabling the scanner to self-locate. This eliminates a need to place target markers on the object scanned. Some units can use a mobile phone as an interface; others use a computer tablet (Figure 3.11).

Figure 3.11 Semi-tethered unit connected to tablet for scan visualization

Typical specs:

Working distance range: 0.4–5 m
Volume capture zone: 160,000 cm^3
3D resolution up to: 0.2 mm
3D point accuracy up to: 0.5 mm
Data acquisition speed up to: 220,000 points per second
Color: 24 bit
Light source: integrated LED flash
Data output: Point Cloud format

3.4.3.4   Metrology-level Handheld Scanner

Certain handheld units were designed to meet industry's need to accurately document large-scale parts and machinery. These scanners are designed to work in conjunction with a laser tracker. Laser trackers are metrology grade laser instruments capable of highly accurate dimensions. Laser trackers work by sending a narrow-directed laser beam to a spherically mounted retroreflector (SMR), which directs the beam back to the tracker where the distance is calculated and processed.

The tracker is connected via a cable to the handheld scanner for data returns, and SMRs mounted on the scan device allow the tracker to follow its movements while scanning. The tracker then can establish the positon of the scanner relative to the object being scanned. The handheld scanner is connected to a laptop, which processes and aligns the scan data of the object, aided by the positioning data from the tracker. The laptop displays a real-time view of the scanning's progress, which the user can observe while scanning. A separate cable connection from a battery pack provides power.

Figure 3.12 Handheld laser scanner/laser tracker system being used to scan boat hull

Source: Photo courtesy of General Marine, Inc. www.generalmarine.com.

The unit works by emitting a rapid series of laser "dots" along a straight line across the surface of the object. A camera captures the point's xyz position which in turn is relayed to the computer for image processing. The unit has a built in compensator for adjusting the laser's strength as needed for bright light situations. It can be used both indoors and outdoors. The levels of precision this scanner delivers, combined with its quick scanning capabilities, make it well suited for scanning objects as diverse as small architectural ornament to large ship propellers. Due to its precision and intricate engineering, these units are among the most expensive handheld scanners on the market. When the cost of a laser tracker is added in, the total system can reach costs of $250K or more (Figure 3.12).

Typical specs:

Working distance range: 180 mm
Volume capture zone: up to 60 m
3D point accuracy up to: 0.003 mm
Data acquisition speed up to: 150,000 points per second
Light source: Laser line/structured light
Color: Monochrome grayscale
Data output: Point Cloud format

3.4.4 LIDAR Systems

The acronym LiDAR, Light Image Detection and Ranging, is a remote sensing technology that calculates distance by illuminating a surface with a laser and analyzing the returned reflected light. It is a relatively new technology, although introduced by Hughes aircraft in 1960, the first 3D laser scanner for terrestrial use was invented in the 1980s. Since then, the technology has advanced and matured aided by the growth of computer capabilities and software development. LIDAR has been acquired as a catch-all term for today's diverse array of terrestrial laser scanners.

Terrestrial laser scanners are a subgroup of scanning devices that are primarily used for geospatial measurements. Found in a variety of industries, they are most prevalent in land surveying, civil engineering, architecture, and construction. They provide detailed 3D data of objects and landscapes. These scanning devices are sometimes referred to as a midrange or long-range scanner, 3D scanner, 3D laser scanner, or other similar generic names. All are portable devices which can be set in the field atop a tripod to collect information. The data is collected either internally or on a removable memory (SD) card; it is then downloaded to a computer for processing and alignment.

Today there are a variety of highly accurate portable scanning devices available. Two types of scanners predominate, a time of flight unit and a phase based scanner. These scanners have seen an exponential increase in accuracy, speed, and data collecting capacity over the past 20 years (Figure 3.13).

Figure 3.13 Various tripod-mounted LIDAR terrestrial laser scanners

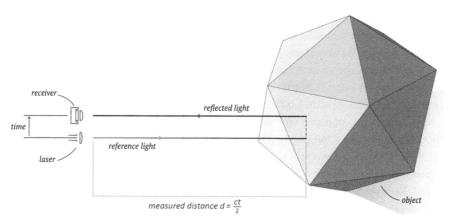

Figure 3.14 Typical time-of-flight laser scanner operation
Source: Image courtesy of Artec3D. www.artec3d.com.

Time of flight scanners are designed to cover long distance environments, both interior and exterior (Figure 3.14). Often used for survey work, certain time of flight scanners will provide precise return on targets and surfaces over 1,000 feet and more. These scanners send a laser pulse from the machine, which is deflected off a beam steering mirror toward the target to be scanned. The scanner, using the known speed of light (c), records the difference in time (t) between the sending of the laser pulse and its return to calculate the distance (d) covered.

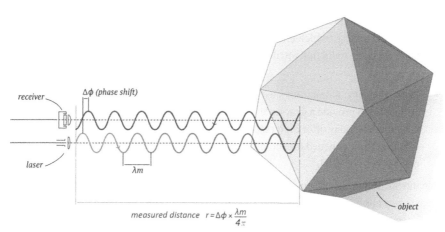

Figure 3.15 Typical phase-based scanner operation
Source: Image courtesy of Artec3D. www.artec3d.com.

Phase-based units are similar in that they use a laser which is redirected by a beam steering mirror to the scanning target (Figure 3.15). However, this laser emits a constant beam (λm) which is set in multiple phases. The phase shift ($\Delta\phi$) of the retuning laser beam is recorded by the systems computer to calculate the distance traveled. Phase-based units have a shorter range than time of flight units, although newer units will provide data returns at distances up to 1,000 feet. Their strength lies in providing a faster scan time and denser array of points returned.

Both scanners have a 360-degree horizontal view, within a +/−270- to-320-degree vertical arc. They record everything within line of sight of the scanner's range. The scanners emit pulses of laser light which capture millions of data points on any surface detected. Each point is positioned in space with an x,y,z designation, creating a 3D digital image referred to as "point clouds."

Manufacturers produce a variety of scanning models. The variations among scanners are designed to meet the specific application needs of diverse industries. Ease of use and cost of the scanner are also design considerations. All Lidar scanners will have the following characteristics:

- Range: Most terrestrial laser scanners will have a point acquisition range that can vary from 50 to 500 meters. The range of the scanner is dependent upon the strength of the laser built into the unit. The laser light itself can be of different wavelengths depending upon the unit's purpose. Some scanners will work better in bright sunlight; others need a dim working environment for best clarity of point accumulation.

- Resolution: As discussed in a previous chapter, resolution is the point density captured by the scanners. This can be a fixed density in some lower cost units, while other models allow ranges in density of low, medium, and high; some offer more advanced settings which allow precise resolution adjustment along with accompanying noise reduction attributes, resulting in finely detailed accurate point cloud data.

- Photo imagery: All scanners use some type of photo capture, at a minimum black and white to provide a grayscale colorization of the point cloud data. Most new scan models today offer high resolution color photographs with the scanning. This function can be turned on and off to allow for quicker, lower data weight scans if color is not required.

Scan controls: When recording data, the scanner must be able to tag each scan's location relative to one another. This is accomplished through a series of different mechanisms built into the operating software of a scanner. Typical controls include:

- IMU sensor: An Inertial Measurement Unit is a device that measures angular rate and force. It is a dynamic sensor that is used by the scanner to calculate its location and movement through a space.

- Tilt compensator: This is a device built into the scanner to determine a parallel horizon plane. Established good survey practice calls for any survey instrument, i.e., the scanner, to be set level on a tripod prior to scanning. Many scanners have an internal level function that displays a level "bubble" to allow for manual adjustment of the scanner. Some scanners have an additional built in tilt compensator that automatically adjusts up to 5 degrees should the scanner be off level.

- GPS locater: GPS stands for Global Positioning System. Most scanner today are equipped with a built-in GPS locator that can be switched on or off when scanning. If the GPS switched is enabled, the scans are tied into the global satellite network to provide longitude and latitude data for the scans. The GPS locator will not work when scanning indoors.

- Altimeter: The altimeter measures altitude, or height, of the scanner relative to a fixed zero elevation. When used in conjunction with a GPS device, it will locate the center of the scanning laser in accordance with established USGS datum level.

- Compass: Many scanners have a compass device built into their system to allow for the scans to be tagged with directional geographic orientation.

There are many manufacturers of LIDAR scanners, as well as multiple scanners made by each manufacturer. The manufacturer's various models will vary slightly in their preset internal precision levels, ability to set scanning parameters, and the time required to complete one scan. No matter which unit is chosen, most terrestrial laser scanners are highly precise, sensitive computing devices that are capable of producing extremely accurate scanning data when proper scanning methodologies are followed.

3.4.5 SLAM

SLAM is an acronym for a technology known as Simultaneous Localization and Mapping. It refers to a class of devices that have the ability to continuously collect data while also tracking the location of the device as it moves about a space. It is often referred to as mobile mapping. Some of these devices rely on technology known as vSLAM, or visual SLAM. They create 3D mapping using photography only. Other devices rely on a LiDAR sensing unit to generate the mapping. SLAM scanners can be handheld, manually pushed in a wheeled cart, or set on a drone or mobile device that is remotely operated.

Conventional 3D laser scanners are differentiated by the technology designed into their devices, such as the laser type and strength and the ability to control the density and distance of the points collected. SLAM laser systems are very similar in their hardware technology, relying on two basic components. They all use a surround view compact LiDAR puck along with an IMU (Inertial Measurement Unit). This assembly enables capture of the data, via the LiDAR Puck, and localization, via the IMU.

Lidar pucks are an enclosed, compact laser sensing unit that operates on a time-of-flight basis. Unlike traditional scanners with rotating mirrors, pucks appear as a stationary unit with no moving parts. They have multiple lasers installed in them, which enable a 360-degree horizontal field of view and a 30-degree vertical field of view. Most units have a 100 meter range with a precision of +/−3 centimeters. They can also capture 300,000 points per second. Often dual LiDAR sensors are used to capture data at varying heights simultaneously.

What primarily separates SLAM scanners from one another are two things:

1. The method of mobility used by the scanning devices.

2. Their proprietary SLAM software, which uses unique algorithms to enable the mapping and LiDAR data processing.

Commonly used SLAM scanning devices are typically a pushcart mounted, handheld, or wearable device to enable quick movement through and around a space. They employ LIDAR scanners that utilize SLAM technology to map indoors and outdoors. A handheld unit is lightweight, designed with multiple lasers set inside a rotating sensor capable of capturing 300,000 points per second at a range of 100 meters with a relative accuracy of 6 millimeters. While designed to be handheld, it can be attached in backpack-like fashion and walked through a space (Figure 3.16). The SLAM device can also be mounted on a vehicle or drone to cover larger areas. Many have an option for an HD camera attachment, which provides high resolution color photographs to be digitally aligned with the scanning data. Once the data is collected, it is downloaded into specialized software to process and align all the data into a complete 3D digital image of the area scanned.

Figure 3.16 Typical wearable SLAM device self-contained within a backpack assembly, designed to capture point cloud data as one walks through a space
Source: Image courtesy of GeoSLAM, www.geoslam.com.

One type of a wearable SLAM device uses two LIDAR pucks combined with a lightweight frame structure that rests on the shoulders. One puck is mounted to sit above the head, another at chest level. It includes a built-in screen so the user can see the scan imagery unfolding as the device moves through the space. Atop the unit are four high resolution cameras that enable a clear field of view for 360 degrees. Like the handheld unit it will scan continuously as one moves through the space. It can record up to 600,000 points per second, yielding a dense well-defined point cloud deliverable. For added alignment precision, it can take a 360-degree panoramic scan at set intervals. The units also have the ability to be record survey control targets. When properly scanned using controls the resulting digital data can reach survey grade of 3 millimeter accuracy levels.

The resulting data from most SLAM devices is compatible with typical scan data file formats, allowing it to be imported in a variety of third-party CAD and modeling software packages (Reference Chapter 9). These devices have been a game changer in acquiring geospatial data. By enabling rapid movement through a space with a scanning device they have eliminated the need for multiple static station setups, thus saving time in documenting large building interiors. Its field survey pace can be up to ten times faster than standard static terrestrial scanning. While a drawback to SLAM technology is its decreased level of precision when compared to static scanners, the overall accuracy level is often well within the parameters required for a wide variety of documentation needs. They are especially useful in creating a digital twin of large industrial space when millimeter level accuracy is not required. Moving rapidly through a space they can document a large facility within a day, usually while the facility maintains its operations.

SLAM technology continues to mature. Some models of SLAM devices are now able to capture and use survey controls in their datasets. Their updated algorithms and filters allow for cleaner data with high resolution visuals, all in a measurable digital format. This marked increase in accuracy, graphic clarity, and documentation speed is making SLAM-based scanners a valuable tool for industrial engineering, architecture, and construction.

3.4.6 Photogrammetry

Photogrammetry is a term applied to a process that uses multiple photographs taken with a standard DSLR or point and shoot camera to produce a measureable drawing or image. While not technically a laser-operated technology, it will often produce a 3D point cloud data deliverable using 2D photographs. The photographs must be taken from various locations and provide significant overlap of the object to be documented. When properly photographed and processed within specialized software, the resulting digital data will provide useable measurement data along with 2D and 3D image information. By leveraging the visual clarity, ease and wide-ranging image capture area of photographs, photogrammetry is an easy method to gain data on large areas of buildings or terrain that would be time consuming or extremely difficult to scan using terrestrial laser scanners.

There are two main categories of photogrammetry, aerial photogrammetry and ground-based close-range photogrammetry. Both use a similar approach in data acquisition method and processing.

Once photographs have been taken, the key to obtaining photogrammetric data relies upon the abilities of the software. There are many software packages on the market today, many of which are listed in the software reference section of Chapter 9. Some advanced software packages with expanded capability can be costly and require training and practice to effectively use them. Fortunately, there are many companies providing software training with experienced instructors and resources to aid through the learning process. For those just entering into 3D scanning or for simple applications, there are free downloadable programs. These downloads enable experimentation with photogrammetry and exploration of its potential without a large investment of money.

Photogrammetric software uses a few algorithmic-based techniques to process multiple photographs into a complete orthomosaic image. Two of the most common are Multiview Stereo (MVS) and Structure from Motion (SfM).

Multiview Stereo is a technique that uses stereo correspondences of identical features between two or more photographs to establish the most likely 3D shape that produced those photos. At its simplest form, it relies upon a series of overlapping photos as well as camera parameters, such as lens focal length or approximated location and orientation, to construct a 3D image. 3D images of the photographed objects are constructed based upon visual information only. The resulting object has no scale; if one is needed it must be based upon a known size of an object within the image produced. Using that object's known size, the image can be enlarged or reduced to match a desired scale. MSV algorithms are limited in their capability. Many software suites use them for initial object formation, but add in structure from motion algorithms to provide computation of camera location parameters and creation of a well-defined and scaled final image.

Structure from motion is an algorithmic-based process used within specialized photogrammetry software to create a 3D model of topography or objects from overlapping 2D photographs taken at various locations and orientations. SfM coding works by identifying common features among a series of photographs. By analyzing the feature changes in imagery and scale the software establishes a spatial relationship between images, based on a local geo reference scale. The software then takes the photos and extracts them as a dense point cloud or mesh, blending them together and adding textures to form the desired image.

Photogrammetric data acquisition is often faster and uses less expensive equipment than LIDAR. However, processing time can often take days when large quantities of photographs must be compiled. A standard DSLR camera is all that is needed in the field. It can be handheld, or tripod mounted; attached to a large pole, balloon or drone as needed to take the photographs. Targets or measuring sticks can be incorporated into the photos; this enables scaling the image to obtain accurate dimensions from the resulting model or map. For best results, images should be taken at varying distance and locations from the object to be documented. This enables the feature details of close-up photos to be used to help align pictures taken farther away.

Processed data can be exported in a variety of file types, including typical 3D software formats such as 3ds, igs, obj, stl, and txt among others. This allows for importation of the 3D object into programs for development of CAD models, animations, or further study. As computer capacity increases and digital cameras develop higher resolutions and add-ons such as GPS locaters, photogrammetry programs will evolve to promote widespread use of this low-cost alternative solution for 3D data capture.

The key to any dependable photogrammetry software is how well it matches features in one photograph to features in another, or, in the terminology of the trade, how well it solves the correspondence problem. The problem can be daunting because the same feature may look very different depending upon the angle at which a photograph is taken, or if the lighting on the object changes from one photo to the next.

One way that metrologists can help the software solve the correspondence problem is through the use of targets. The targets are liberally applied on or around the object before the photos are taken. They provide the software with several easily identifiable points to check for correspondence. The simplest form of target is a highly reflective circle surrounded by a black ring. Another common form of target is a strip of black tape on which a series of circular reflective spots are printed. Targets can even be black stickers with reflective individually coded shapes. The unique reflective shapes help prevent the software from misidentifying a target, an unfortunately common occurrence with circular targets. Targets come in different sizes to suit different sizes of objects, and ranges of camera shots. The tape strip targets come in a variety of circle-to-circle distances that can help distinguish one line of target tape from another in a photo.

Camera quality is important when doing metrology grade work. Although photogrammetry software can work with digital photos taken with a cell phone, commercial cameras built for metrology are more exhaustively checked for defects in their lenses. Their frames are engineered so that they will remain stable if environmental conditions change.

REFERENCES

Artec3d. https://www.artec3d.com/portable-3d-scanners/artec-leo. (accessed 5 August 2022).

Creaform3d. https://www.creaform3d.com/en/portable-3d-scanner-handyscan-3d. (accessed 5 August 2022).

Faro USA. https://www.faro.com/en/Products/Hardware/Freestyle-2-Handheld-Scanner; https://www.faro.com/en/Products/Hardware/Quantum-FaroArms. (accessed 5 August 2022).

Geoslam. https://geoslam.com/us/product. (accessed 5 August 2022).

Hexagon. https://www.hexagonmi.com/en-US/products/portable-measuring-arms; https://www.hexagonmi.com/en-us/products/laser-tracker-systems/leica-absolute-tracker-at960. (accessed 5 August 2022).

Lazar, D. (2020). An Explainer on Structured Light vs Lidar for 3d Depth. http://www.crowdsupply.com/onion/tau-lidar-camera/updates/an-explainer-on-structured-light-vs-lidar-for-3d-depth. (accessed 10 November 2021).

Martinez, L. (2022). Aerospace companies flying high with 3D laser trackers. http://www.qualitymag.com/articles/96912-aerospace-companies-flying-high-with-3d-laser-trackers *Quality Magazine*. (accessed 12 August 2022).

Navvis. https://www.navvis.com/vlx. (accessed 5 August 2022).

Chapter 4

Data Acquisition and Processing Software

4.1 INTRODUCTION

Given the speed and complexity of acquiring scan data, it's inevitable that some kind of computerized control interface is required. On some instruments, such as long-range scanners, the interface is built into the scanners' on-board computer. Most scanning instruments however rely on an interface through a connected computer. No matter how the interface occurs, to make it work specialized software is required.

This software can be classified into three functional categories:

- Data Acquisition – collecting and processing point cloud data

- Inspection – comparing the point cloud data to the nominal, or design intent

- Modeling – also called documentation or reverse engineering, creating a visual representation (via CAD, or physical 3D model) of the data.

Often two or more of these functions are combined in a single software package for practical purposes and usability.

3D Scanning for Advanced Manufacturing, Design, and Construction, First Edition.
Gary C. Confalone, John Smits, and Thomas Kinnare
© 2023 John Wiley & Sons, Inc. Published 2023 by John Wiley & Sons, Inc.

4.2   DATA ACQUISITION SOFTWARE

Data acquisition software can be obtained from an instrument manufacturer, a third-party developer, or one's self-developed proprietary scripts and algorithms. Software written by the instrument's manufacturer might only function with a particular brand of instrument. Third-party software, written by an independent company, is typically compatible with a number of instruments from different manufacturers, often even serving different instrument types.

Creating one's own data collection interface requires programming skills. Some manufacturers have written Software Development Kits (SDKs) for their equipment which assists in creating an interface. The kits effectively give one access to the basic instrument commands. The developer then modifies and augments script routines as needed to serve their specific purpose. Creation of an interface using an SDK can be a very effective way to automate simple repetitive jobs.

Regardless of whether you are using an on-board interface, commercially available software, or a custom self-built interface, there are various features that the data collection software should provide in order to serve its purpose. These features, some required, others desirable, help to minimize the time to obtain calculations while maintaining a level of accuracy and consistency in the results.

4.2.1   Instrument Compatibility

One very desirable feature with data collection software is compatibility with many instruments types and manufacturers. Data collection software typically costs between 25% and 50% of an instrument's price. It's a substantial investment that should not have to occur every time a new instrument is purchased. Software from instrument manufacturers generally works with multiple models of their own equipment, but usually not a competitor's.

Third-party software is written by companies without direct ties to equipment manufacturers. These companies contract for access to the manufacturers' SDKs. Their software usually works with a number of the most common brands of equipment. As an added advantage, the third-party software usually incorporates post-processing, analysis, or reverse engineering functions. These added functions are often the primary reason the software is purchased. Even if third-party software is not compatible with a certain instrument, such software usually has the ability to import and read data in a wide variety of formats.

Spatial Analyzer, Polyworks, Verisurf, and Geomagic are examples of third-party data acquisition software that functions with a wide variety of instruments. The downside to third-party software is that there may be occasions where common operations like instrument checks, compensation, settings, and even measurement procedures are more awkward, or less streamlined than found when using the instrument manufacturer's own proprietary software.

4.2.2 User Interface

A simple user interface is probably the most desirable feature. Software that is difficult to learn and use makes processing time intensive and prone to mistakes. Common commands should be readily accessible with a minimum of mouse clicks; they should not be hidden in layers of menus. Some of the most important commands that should be easily available are those for view control, measurement, feature and point cloud visibility, point cloud editing, feature creation and editing, and settings.

4.2.3 View Control

View control must encompass three operations: zoom, rotation, and scroll, often called pan. It can also include clipping planes or boxes. Clipping boxes define a volume within which point clouds are visible; anything outside the box is not shown. Likewise clipping planes can be used to section a point cloud and show only the portion above or below, left or right, in front of or behind a defined plane. This can be a great help in understanding what a point cloud represents. Point clouds are often portrayed as translucent, and it can be difficult to distinguish the depth of points in the computer screen's field of view.

Some software allows establishment of a section plane. The software will then highlight all the points within a given distance of that plane. The distance is usually defined by the user. It is in many ways similar to a narrow clipping box but doesn't hide portions of the remaining point cloud. This visualization can be very helpful in determining what the point cloud is actually showing.

4.2.4 Point Cloud or Mesh Visibility

Data collection software converts the x,y,z data numbers derived from the instrument into the visual format of a point cloud or structured mesh. The ability to view the data within that visual interface is very important. If shown as a point cloud, can the point cloud be given grayscale values or colorized? Can it be shaded by surface normal and light direction? Can the size of the points and number of points displayed be adjusted? If so it will look more like a real object, and help one to decipher parts and pieces of the object, thus making it easier to analyze and evaluate.

Structured meshes, also known as polygon models, cells, or STL models, attempt to represent a point cloud as a segmented surface. The most common method uses a net of flat triangular surfaces to represent and define the part's surface. By creating a solid surface of mesh sections, the amorphous quality of a point cloud is removed, and the part takes on greater visual clarity. The object becomes much easier to visualize based on a mesh representation.

There should be a thinning function for data display. A thinning option reduces the number of points visible in the graphics window. This can improve a computer's response time by minimizing graphic card response time trying to represent millions of points.

Sometimes the thinning is done automatically by an algorithm in the program, some software allows for manual setting adjustments.

Oftentimes there are two thinning settings, a dynamic and a static one. The static setting is applied when the view is not changing and displays the full amount of points or mesh. Because the view isn't changing, the program isn't trying to recalculate point locations on the screen. The dynamic setting is applied when you are manipulating the view by zooming, scrolling, or rotating. The dynamic setting reduces the number of points shown, hence minimizing the number of recalculations required by each move. This in turn lightens the load on the graphics card and improves the visual response time of the screen view.

The software should have the capability to display or hide point clouds or meshes individually or within selected areas. More advanced software is able to show or hide multiple areas of selected portions of the point clouds or meshes. A useful capability is for the software to remember different selected regions of point clouds and provide control visibility based on region name.

4.2.5   Measurement and Scanning Parameters

Any data acquisition software must have measurement functionality. These are controls to start and stop scans, and to set scanning parameters. These controls should be easily accessible. Scanning parameters can differ dependent upon the type of equipment. Some of the most common parameters are scan density, resolution, daylight exposure and filters such as return intensity or noise reduction. The ability to restrict and define scan boundaries has value as it prevents measurements outside the area of interest and can lessen scan time in the field.

4.2.6   Point Cloud Editing

Basic point cloud editing functions are a desirable component of data acquisition software. The data can also be seen on a view pad and checked while scanning to insure a complete capture of the surface is made. While scanning one will see the object as well as any other parts or pieces that may not be required. That presents an opportunity to eliminate extraneous data. Anything done to minimize the dataset prior to post-processing will make that stage of the process smoother and faster. During data acquisition the most important point cloud editing functions are deleting point cloud points, point cloud thinning (sometimes referred to as sampling or filtering), and noise reduction. When working with mesh data, decimation will reduce mesh count, and smooth surfaces, achieving a result similar to point cloud thinning.

4.2.7   Post-processing Functions

Functions often associated with post-processing can also be advantageous in data acquisition software. Creation of basic geometric features like planes, cylinders, or surfaces from the measured data can help during the post-processing of the data by providing a visual frame of reference or easily identifiable landmarks. If geometric features have been created from the point cloud data, the software should be able to show or hide those features individually or by type.

The ability to either create or integrate a coordinate system within the raw data is often a highly desired feature. By using the coordinate system, the point cloud data can be aligned to a desired reference frame before post-processing begins.

4.2.8   Settings

Settings will determine all of the instrument parameters, functions, and default values. They are often placed within a menu tab or an obscure corner of the interface. This is primarily due to the limited access this function requires. Once set, they are seldom changed. However, it is important to understand where the settings function lies in the interface, what it controls, and how to adjust as needed based upon the project's requirements.

4.2.9   Graphics Window

The aforementioned set of functions typically require a toolbar or icon button for each one, some with multiple options. All these buttons and toolbars are at odds with two basic desirable features, a large working screen and a friendly user interface.

A large graphics window does not refer to the size of the monitor. Whatever the size of a monitor, the program window has a limited area. In a poorly designed interface too much of that area is consumed by toolbars, ribbons, buttons, feature trees, and status windows. The result is that the graphics window, the visual working area that displays the point cloud data is smaller than is truly convenient. Sometimes it is not even understandable without incessant scrolling and zooming in and out.

Most programs use Windows Toolbars or Ribbons as the basis of their interface. It provides a familiar looking environment for the user. A good design will provide a large graphics display window without burying commonly used functions behind too many mouse clicks.

4.2.10   Instrument Location

Many long-range scanners, especially those suited for architectural or large-scale site work, have some means of estimating their position and orientation. Different methods are GPS, gravity levels, or photographic monitoring of the position as it moves from scan location to scan location. This kind of functionality can be used to approximate actual scan locations and register scans during the scanning process. The pre-alignment of the scan data enables the final registration process to be resolved quickly and with more accuracy.

Instrument-specific software will always be able to take advantage of this capability. Third-party software usually will not.

4.2.11 Scan Bounding

Often it's desirable to limit the extents of a scan. This occurs most often in architectural scanning when a high-density scan is desired, but only for a small area or section within the scanner's total field of capture. Considerable time can be saved by limiting the horizontal sweep of the scan. Whatever percentage of the horizontal sweep you remove will provide a corresponding percentage in scan time loss. Some interfaces allow for bounding boxes to be placed around on the object, thus affecting the vertical sweep of the scanner as well. This results in not only time savings but saves large quantities of unneeded data from becoming embedded in the dataset.

Another form of scan bounding is to limit the maximum range of the scanner. Points measured beyond the limit are discarded. This can improve quality because closer points are usually more accurate than points very far away. It can also improve downstream processing time because the loss of points outside the region of interest reduces the point cloud data size, enabling faster processing within the computer. This technique is most useful in architectural scanning.

4.2.12 Acquired Data and Display

Another nice, but not essential, feature is the ability to display the point cloud in a variety of formats: real color, intensity, or shaded. Real color data is usually acquired by cameras and mapped onto the point cloud. Intensity-based displays are color coded based on how strong the return of the laser was for each point. Usually, red indicates a strong return and blue tones indicate a weaker one. Shaded displays, even if monochromatic, can look very realistic, and confirm that the instrument is collecting data on the normal in addition to the x,y,z location of each point. This function is very important for long-range laser scanners.

4.2.13 Quick Measurements and Point Interrogation

It's useful to be able to field check dimensions. Being able to easily and quickly check point to point distances or point coordinates can be a big help in verifying the accuracy of the scanner and resultant data.

4.2.14 Compensation and Verification

These functions usually apply to smaller scanners for mechanical inspection or reverse engineering. They provide a rapid means to verify that the instrument is working properly, or if adjustments to the instrument to accommodate current conditions are required.

4.2.15 Geometric Primitives

It can be useful to associate point data to geometric entities such as planes, lines, circles, spheres, or cylinders while in the field to facilitate alignments or registrations. It's an added bonus if these geometric features can be fixed in diameter, or orientation. This kind of functionality is usually found in mechanical inspection or reverse engineering software.

4.2.16 Point Identification

Another nice feature is to be able to flag points and give them special status and a name. These points can be used for alignment to a coordinate system or as seed points for cloud registration.

4.2.17 Control Point or Feature Import

Similarly, it can be valuable to be able to import points or geometric features from CAD or a data table. These points can then be applied to the data when scanned, and assist in preregistration where available.

4.3 INSPECTION SOFTWARE

One of the most common and useful analyses done using point cloud data is to compare measurements from an item with the ideal or prescribed shape. If CAD data defining the ideal part is available, the process is usually quite simple for the user. After aligning the data to the CAD model, the user selects the measured points or region of mesh to be used in the comparison, and then selects the corresponding CAD surfaces with which the measured data are to be compared. The inspection software calculates the shortest distance from every selected point to the surfaces of interest. Results are most often displayed as a colored puddle lot, or as a whisker plot. Puddle plots, called heat maps by some, are topographical maps that use colors to show how far measured points are from the reference surface.

Another way to chart deviations is using a whisker plot. They are an older method of representation, and one better suited to sparse data. Whisker plots use short line segments to illustrate the deviations between measured points and the reference surface. The length of the line segments is proportional to the deviation. The scale of line length to deviation can usually be set as low as 1 to 1, but for large parts, or parts with small deviations the scale can be 1,000 to 1, or even higher. Often these plots use one color to show points that are higher than the reference surface, and another color to show points that are lower. Some software allows the user to combine puddle and whisker plots. The whiskers change both length and color depending on the magnitude of the deviation they represent.

No matter which graphical representation used, if comparing multiple parts or different views of the same part, it is good practice to use a consistent scale throughout. This ensures that whomever reads the report will be able to compare and contrast the various objects for accuracy.

4.4 DATA EXPORT OPTIONS

Clients often have varied workflows and require dataset deliverables to be in a specific file format. Many programs used for registration, analysis, inspection, or reverse engineering only accept a few point cloud formats, although they may export files in a wider variety of formats. It's advantageous to have a software that can export and import multiple formats. Some formats can only be used within a company's proprietary software. Some formats carry all the data a scanner acquires; some carry only x,y,z data for the points.

File compression is an important consideration. It sometimes affects the data quality, but always affects the data weight of a file. Some formats have no compression, making the files heavy in gigabyte weight. Others are highly compressed, and consequently much smaller in data weight. All of these factors will place restrictions or requirements on the final data file format required for a project. Proprietary formats, those from the instrument's manufacturer, tend to have compression as a built in feature. Therefore, they tend to produce smaller datasets. They typically also include color, intensity, and normal data.

A few of the more widely used nonproprietary formats are:

- E57 – A point cloud standard established by the data interoperability subcommittee of the ASTM E57 Committee on 3D Imaging Systems. It is a vendor-neutral file format for point cloud storage, allows one to store images and metadata produced by laser scanners and other 3D imaging systems. E57 is typically a compressed binary file format.

- XYZ – These are ASCII database files which contain numerical representations of the various points collected in a point cloud file. No compression.

- STL – Describes point cloud as a structured mesh of triangular planes. Typically a large file format, typical for most mesh file formats.

- IGS, IGES – Typically a computer-aided design (CAD) file format. It typically describes a surface model. Can carry data on other geometric features in addition to points. Larger file sizes.

- STEP – Newer technology than IGES, often used to describe datasets of solid models.

4.5 POINT CLOUDS

4.5.1 Introduction

The typical dataset generated by most 3D scanning equipment is referred to as a point cloud. The name is a reference to its imagery when viewed in software; a 3D image composed of a multitude of small dots with a cloudlike opacity floating in the computer screen. Each dot represents the x,y,z coordinate of a distinct point on a scanned surface, given either a grayscale or RGB (colorized red-green-blue) value for ease of visual recognition. One scan will often contain in excess of a million points. By combining multiple scans taken from various viewpoints all sides and edges of any exposed surface can be recorded. When assembled collectively these scan points represent a true scale digital iteration of whatever object or environment was scanned (Figure 4.1).

Point cloud data will vary widely in its quality dependent upon the equipment used, the resolution setting, the object being scanned, the distance and angle of the scanner from that object and the operator's scanning experience. The resultant quality of the data will have a direct correlation to its usefulness in creating an accurate point cloud model.

4.5.2 Point Density

Aside from the use of a high precision scanner that is properly calibrated, point cloud density is the first key element required to create an accurate dataset. This density is often referred to by the term "scan data resolution." Most handheld scanning units

Figure 4.1 3D Point cloud data from multiple aligned scans as viewed within a processing software

will have a preset resolution level, since they are typically always used at a close range, often varying no more than two to three feet in distance from the object being scanned. Mobile or backpack scanners designed to capture point cloud data while walking through a space have settings established by the manufacturer; as the scan data is simultaneously captured, processed, and aligned within their proprietary software. However, when using a terrestrial laser range scanner, it is up to the user to establish scanner settings, as well as procedures for downloading, processing an alignment of the scan data. One important component, the scan density, or *resolution*, must be set by the operator. Understanding how a scanner compiles a point cloud dataset is key to knowing how to set and obtain proper scan density levels. Key elements to consider:

- Resolution level settings
- Surface characteristics of object to be scanned
- Distance from scanner to object
- Angle of incidence
- Scan overlap
- Time available for scanning on-site

4.5.3 Resolution Levels

Terrestrial laser scanners, sometimes referred to as ground-based long-range scanners, are typically used in surveying buildings or medium- to large-sized objects. Two operational types are common, a phase-based scanner and a time of flight scanner. A phase-based unit emits a continuous stream of laser beam that is put in multiple phases. Each point is derived from a measurement based upon the change in phase waves detected by the scanner. A time-of-flight unit shoots a series of laser pulses, measuring the time it takes to reach the object and return to the scanner. In simplified terms, both emit a rapid series of laser light that slowly spread apart and paint the surface of any object encountered. Similar to a flashlight shining on a wall, the closer the object to the scanner, the denser and more concentrated the circle of light (Figure 4.2). By increasing the resolution level (i.e. scan density) of the scanner, the pulses are more frequent, thus tighter spaced, creating a denser placement of data on the scanned object. By achieving a closely packed set of data points on a surface, precise measurements can be taken of slight surface undulations or rifts. The amount of detail to be captured on the object to be scanned will be a prime factor in determining what resolution level is best to achieve optimum data coverage. When scanning flat surfaces such as walls and floors, the scan data needn't be as dense since there is little surface deviation. However, when scanning organic objects or highly detailed surfaces a dense dataset taken at high resolution is needed to capture the minute measurement variations between points on the surface scanned.

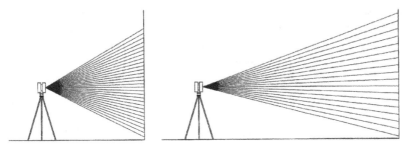

Figure 4.2 As the scanner is set closer to the object to be scanned, the density of scan points on the surface will increase

4.5.4 Surface Characteristics of Object to be Scanned

In order to obtain scan data, the scanner must be able to collect and record the return of a laser pulse from an object. The lapse in time between the emitted and received laser pulse is what enables the collection of precise x,y,z coordinates which indicates the location of the surface relative to the scanner. Some surface properties, such as color and finish, will be better than others in returning light pulses. Bright sunlight can also diminish the scanners ability to collect accurate data by drowning out the return signal. Matte surfaces in a neutral color range that are not subject to harsh sunlight or high artificial light levels will produce the best results in returning scan point data. Scanners typically yield very good pulse return counts when scanning in areas of low light levels such as attics and basements. Likewise scanning outdoors in cloudy overcast conditions, or at dusk or dawn will yield better results than scanning midday under bright sunlight.

Scanning reflective surfaces such as glazing, mirrors, and highly polished stone and metals can be problematic. Transparent glass reflects little or no light, thus offering no data capture of the window's surface location. It will however allow the light to pass through and return back to the scanner locating the objects behind the glazing. Mirrors and mirrored glazing are a different story; they reflect light from the mirror's surface equally which forms a mirror image of scan data. When viewing that image, it appears to be located on the opposite side of the mirror from the space or roomed being scanned. When scanning mirrors or highly reflective objects a light powdery coating is often sprayed on the surface. This enables the surface to yield a proper range finding feedback of the laser pulses.

4.5.5 Scanner Distance and Angle of Incidence

Aside from the resolution setting, placement of the scanner in relation to the object being scanned has the most important impact on the resulting scan density. Objects that are close to the scanner location can be scanned at a lower level of resolution than objects further distant. The further away an object is from the scanner, the points striking its surface will be more widely spread out and lower in number. However, the wide range of laser pulse rays emitting from the scanner will land on a surface in in a variety of angles. The deviation from a perfect 90-degree pulse is referred to as the angle of incidence. The closer the angle of incidence is to 90 degrees, the higher the likelihood of that laser point returning to the scanner, resulting in a higher density of surveyed points from the scanned surface. As the angle of incidence increases either above or below 90 degrees (Figure 4.3) the resulting scan data will diminish, lowering the resulting scan density achieved from that surface. This change in scan angle can be attributed to both the location of the scanner in relation to the object scanned as well as the physical angular attributes of the objects surface. For example, scanning the side wall of a building with a gable roof will collect fairly dense data from the wall façades which are perpendicular to the scanner, but the roof will lose data resolution due to the pitch of the roof which produces an increase in the angle of incidence, causing the majority of scan laser points to either spread apart or bounce off away from the roof instead of returning back to the scanner for capture (Figure 4.4).

4.5.6 Scan Spacing and Overlap

Scanning is a line of sight process. As such, the spacing and location of the scanning positions is very important. In order to adequately document an object or space one must scan from various view angles. Most scanning projects require multiple scans, which result in significant overlap of surfaces scanned from different angles (Figure 4.5). Once

Figure 4.3 A view of a single point cloud scan on a roadway, indicated by the open circle, in section view and plan view. Note how the density of the scans is reduced as one moves away from the center of the scan increasing the angle of incidence

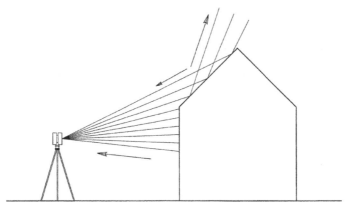

Figure 4.4 As the angle of incidence increases, two things occur: (1) the scan points begin to spread farther apart on the surface, reducing density and (2) the increased scan angle reflects a larger portion of the laser light away from the scanner, leading to a reduction in data points recorded by the scanner's receiver

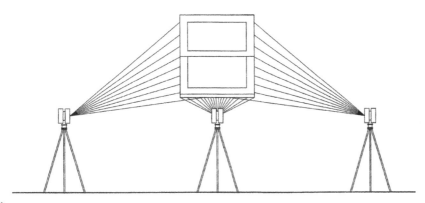

Figure 4.5 Scanning requires line of sight; multiple scans are required to adequately record all edges of the object being scanned

aligned, surfaces will have scan data from multiple scans, which increase the level of detail for that object. Scans taken closer together will create more overlap in data from one scan to another, adding to the overall final level of detail on the object.

4.5.7 Time Available for Scanning

A final factor to consider when setting resolution is the total number of scans needed for the project and the time available for scanning. As the resolution increases, the time needed to complete a scan grows geometrically, with a corresponding increase in the [megabyte] size of the scan. If scanned at too high a resolution, more time is required in field work. The resultant dataset will also be very large, possibly hindering computer performance when CAD modeling with point cloud data. However, scanning at low resolution levels to save some field time can yield a scan dataset lacking in the imagery needed for accurate CAD modeling. Most projects are scanned at various resolution levels, dictated by the size and configuration of the areas being scanned, to enable adequate data for preparation of the required deliverable.

Most scanning equipment manufacturers will have suggested settings for the use of their scanners. Using those as a basis and adding hands-on field experience and experimentation with the scanner at various resolution levels will enable one to understand how to adjust those settings to provide an accurate, quality point cloud ready for processing.

4.6 PROCESSING AND REGISTRATION

Once individual scans have been acquired, they are downloaded into a computer and ready for the processing stage. Processing of the data takes place within the scanning hardware manufacturer's software or other compatible specialized processing software. The first stage is preprocessing of the raw scan data. This is an automated process, usually within the equipment manufacture's software, that removes noise and extraneous data points that could interfere with data analysis during registration. Raw scan data may contain various scanning artifacts. These can include surface noise, data point outliers, low data return on fine surface details, or holes in data from visual occlusion or reflectiveness of surfaces. The raw data will also include many redundant scan data points caused by the overlap from multiple scan viewpoints. Initial preprocessing will run each scan through algorithms to clean up the scan data of outliers and noise. It also creates a color or black and white image of the scan by assigning an RGB or Grayscale color to each individual point within the data file. The resulting image allows one to see the detail and clarity of the scan image. It is now ready for further processing and proper alignment of the scans, referred to as scan registration.

The process of scan registration involves accurately aligning multiple individual scans together to form a digital 3D image of the area or object scanned with one common based coordinate system. This process is sometimes referred to as "stitching," likened to sewing of a shirt or dress from multiple small pieces of fabric. This process can be done manually or automatically, dependent upon the software used and the user's preference. Most scan alignment software will provide both "Target" and "Cloud to Cloud" registration techniques. A user can work with one or a combination of both methodologies to arrive at a comprehensive, accurate aligned point cloud dataset. A comparison of the two methodologies is as follows:

Target registration: When scanning an object or space, multiple targets are set in place. Each scan should have at least three targets in common with an adjacent scan. The targets are typically either checkerboard paper targets set on walls, floors or ceilings; or spheres placed at random heights and intervals within the scan field. The registration software will recognize these targets, via manual identification or an automated process. By using triangulation between the various targets, the scans are aligned in space relative to one another. In addition to targets, planar objects or specific points common to multiple scans can be identified and used in association with the targets to align scans. When done properly target registration can be extremely accurate (Figure 4.6).

Cloud to cloud: The process of aligning one scan to another by matching common overlapping feature objects (for example walls or floors). In order for it to be effective there must be sufficient similarity in the geometry of overlapping areas. This method has become very common in most software on the market today. One of its key advantages is that it eliminates the need to use targets when scanning. Cloud to cloud scan registration relies on the scanned objects features to provide a means of commonality between scans. When scanning spaces with a high level of redundant features the software can sometimes mismatch objects, skewing the alignment. When this happens the software user will have to manually adjust the scans position, or identify overlapping common targets between scans to assist the software in proper alignment via feature recognition.

No two scans will align 100%. While deviations of +/−0.0026′ between two adjacent scans is common and acceptable this small discrepancy in alignment will grow as multiple scans are placed together. This is referred to as scan drift. In scans of small spaces or objects drift has very little consequence on the data's overall level of accuracy.

Figure 4.6 A LIDAR scanner with three common targets used when scanning on-site: A swivel-mounted checkerboard monopod for control points, a wall-mounted paper checkerboard target, and a target sphere

However, when scanning large spaces or open areas, the drift accumulated by numerous consecutive scans will cause significant measurement distortions across the overall project area. This can be eliminated by applying a set of survey controls to your scanning registration.

4.7 SURVEY CONTROLS

Once scans are properly aligned and registered, they are all bound by a common coordinate system. This is what allows one to interrogate the scan data and derive accurate CAD models and surface deformation information. The coordinate system can be one derived by the on-site scanner settings, or it can be a coordinate system applied during the registration process to match specific site-established control points or USGS datum.

In order to utilize controls, a system of targets and points must be established in the field, typically by a surveyor or technician using a total station or laser tracker equipment. Referred to as survey controls, these station points can be placed on-site, intended for either temporary or permanent status. Typically during construction, a set of site controls will be established so that all trades are working within the same geodetic constraints. In a large factory or warehouse, permanent controls established will insure that updated 3D scans of the facility will be tied to the original CAD and scan documentation, allowing ease of integration for new installations within the facility.

Controls are a tabular list of named or numbered data points in x,y,z format, referred to as a .csv (Comma Separated Values) file (Figure 4.7). At the completion of the registration process, after the software has aligned all the scans, the .csv file is brought into the software and applied to the scan project. The software automatically aligns all scan data to the x,y,z values found within the .csv file. If a proper set of controls has been established in the field, all aligned scans will rotate and reshape conforming to those coordinates. This will minimize drift inherent in large and multilevel scan projects, as well as set all scan points to a datum coordinate system compatible to the project's intent and design.

Ref Point	Y	X	Z
G103	4916.173	5188.238	24.387
CB336	4929.251	5164.797	24.985
CB337	4929.015	5157.015	25.031
CB338	4902.952	5155.395	23.008
CB339	4916.654	5122.39	23.126
CB340	4913.628	5084.231	22.218
CB341	4902.49	5044.613	23.134
CB309	4935.36	5105.127	23.561
CB306	4936.381	5121.54	25.015
CB308	4931.084	5114.848	24.834

Figure 4.7 A portion of a typical .csv file indicating x,y,z values of control points

4.8 POLYGONAL MESH

A polygonal mesh is a file type commonly used in 3D modeling. Graphic modeling and animations are best represented and viewed when objects have a solid surface. While point cloud data provides accurate and comprehensive data, it has two problems in regards to 3D modeling: poor visual quality and the difficulty to alter or amend a point cloud object.

Polygonal mesh enables point cloud data's translucent visuals to become a solid 3D visualization (Figure 4.8). The mesh becomes a solid surface by joining multiple sets of three or more points, or x,y,z vertices, thus creating a series of planes and edges which define an object. Each edge must be part of a polygon; each vertex must be shared by at least two edges. This creates a series of multiple small interconnected surface planes, usually triangles or some form of polygons, all fitting together similar to a jigsaw puzzle. The more polygons used to create the mesh, the better the definition of the object. The smoothness of the surface, and corresponding likeness to the object scanned, is dependent upon the combined accuracy, placement, and number of point cloud vertices used. Many hand-scanning laser devices will provide an output of data already set in polygonal mesh format, typically described as an .stl or .obj file. This mesh data format is used primarily for objects, from small parts to large statues or people.

The advantages of a polygonal mesh file are:

- It provides a solid surface, finely detailed, true scale file format easier for both visualizations and modeling within 3D software.

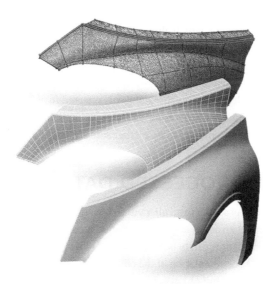

Figure 4.8 Point cloud data to polygonal mesh to solid surface CAD image
Source: Polyworks image courtesy of InnovMetric.

- The polygon mesh is a format which can be easily processed and rendered by most computers.
- Since the solid surface mesh is derived from multiple polygons, it can be adjusted or amended as needed to create computer images of parts and objects, allowing evaluation and study prior to creating the actual part.

There are some issues to be resolved when processing and modeling mesh files:

- File size: In comparison to a point cloud dataset, a mesh file is typically larger, requiring sizable areas of space on a computer's memory and slowing processing. The coding needed to describe the plane surfaces between points requires more data, thus adding considerable weight to the overall dataset.
- Noise: If the polygonal mesh was derived from raw point cloud data, a smooth surface can appear rough or scaly due to uneven relative height variations of the point cloud vertices. This condition is referred to as "noise."
- Voids and holes: This is common to all laser scan datasets of an existing object. Due to overlapping geometry or organic configurations of the object scanned, line of sight obstructions will inhibit the laser scanner's ability to record all parts of the surface. As a result the objects produced will often have voids or "holes" within the raw dataset.

All of these conditions are readily addressed within most 3D processing and modeling software. To reduce a file's size, the number of mesh polygons needs to be decimated. Decimation is done using filters (algorithms within the software), to combine smaller triangles in adjacent planar surfaces, creating a smoother surface with less triangles, thus reducing the file size. Decimation will work throughout the surface area of an object, smoothing both flat and curved surfaces. In addition to decimation, most software provides mesh denoising filters, which eliminate noise spikes and inconsistencies within what should be a smooth surface on the scanned object. Once cleaned of noise and decimated, the 3D object can be adjusted to fill in surface voids. Creating new mesh surface infills is accomplished both automatically and manually by capabilities built into the software. Once completed, the resulting object is referred to as a "water tight" model; a solid surface polygonal mesh file dataset that forms a completely enclosed object with no holes or voids. This file is now ready for further study, modeling, or 3D printing.

4.9 USING PROCESSED DATASETS

Once scan data has been processed and aligned, its benefits can be realized. Its prime benefit is the formation of an accurate full scale digital model which can be shared and studied on a computer. Scanning data is typically used as a basis for three deliverables: forensic study, 3D or 2D documentation of the object scanned, or to create an actual replica of the object, either at full or partial scale. The latter two usually require further work with the data in a CAD or modeling software. However, most scan data software will contain some means for data interpretation and study using the raw data or scan images within the processing software. Whether point cloud or polygonal mesh, the object or structure captured in 3D scans will reveal characteristics often not available by conventional hand survey methods. The object's dataset, now reduced to millions of surface points or planes, can be turned and twisted, cut and sectioned, creating various plan, elevation, and sectional views to reveal surface anomalies and planar alignments.

4.9.1 Measurements

Scan data is a set of x,y,z points, each a distinct measurable point in space. Any processing software will indicate the x,y,z parameters of a point selected on the computer screen. Most allow the x,y,z point values to be reoriented, allowing the operator to establish a specific 0,0,0 origin point for the project, or at a minimum, a zero "z" base elevation. In surveyng and construction z value typically indicates elevation, or vertical height relative to a zero baseline. Scan images are composed of multiple x,y,z points. Moving a mouse pointer across a screen image will display x,y,z values. If the pointer is run along a scan image of a floor the "z" values displayed will fluctuate. These fluctuations indicate either a rise or fall of the floor at that specific point (Figure 4.9). This capability allows one to quickly determine if the floor is level or pitched. A similar method can be used on lines of masonry coursing, conveyor belt runs, or window heads, using the z value of the 3D data to quickly understand changes in level that are occurring.

Figure 4.9 Note x,y,z values displayed when cursor is upon a specific point within the point cloud dataset, as seen within processing software

4.9.2 Cross Sectioning

Most processing software will allow measurement between points and surfaces. By isolating surfaces or sections of the object scanned, distinct dimensions can be taken to check for conformance to design specifications for a surface length or radius. One of the most beneficial study methods is to cut cross sections though the object. Cross sections exist in three types, horizontal, vertical, and angular. Use of these three section views, often taken in multiples along the length or width of the object or structure, will reveal deformities such as tilt or lean, surface depressions, bumps, and ridges. Often capabilities exist that enable a thin measured slice to be taken through all or part of the dataset. That slice can then progressively move up or down along a straight-line plane. This will reveal any movement or sway within the object or structure.

When taken through an interior of a pipe or tunnel, multiple section slices taken at regular intervals will reveal curves or constrictions that occur as one traverses its length. Consecutive section slices of a specified thickness can be cut into a scan dataset of a landscape site or a large floor slab. Starting at a predetermined zero level, slices are cut into the slab or site, generating a topographic map detailing any slope, pitch, depression, or ridges. This is especially useful in determining the flatness of floors in large industrial facilities.

REFERENCES

Hindawi Leihui, L., Wang, R., and Zhang, X. (2021). A tutorial review on point cloud registrations: principle, classification, comparison, and technology challenges. https://www.hindawi.com/journals/mpe/2021/9953910. (accessed 15 December 2021).

IEEE Xplore Shao-Wen Yang and Chieh-Chih Wang. (2008). Dealing with laser scanner failure: mirrors and windows. *2008 IEEE International Conference on Robotics and Automation* 3009–3015. doi: 10.1109/ROBOT.2008.4543667. https://www.ieeexplore.ieee.org/document/4543667. (accessed 3 January 2022).

ResearchGate Xiaoshui, H., Mei, G., and Zhang, J. A comprehensive survey on point cloud registration. https://www.researchgate.net/publication/349787695_A_comprehensive_survey_on_point_cloud_registration. (accessed 15 December 2021).

Thomson, C. (2018). Can surveyors trust targetless registration for point cloud creation? https://www.info.vercator.com/blog/can-surveyors-trust-targetless-registration-for-the-creation-of-point-clouds. (accessed 18 December 2021).

Weyrich, T., Pauly, M., Keiser, R. et al. (2004). Post-processing of scanned 3D surface data. *Eurographics Symposium on Point-Based Graphics.* https://lgg.epfl.ch/publications/2004/weyrich_2004_PPS.pdf. (accessed 4 January 2022).

Chapter 5

Post-Processing

5.1 INTRODUCTION

Since the advent of scanning systems in the 1990s the volume of data acquired by scanning instruments has often been greater than the processing capacity of most computer systems. As the years progressed computer storage capacity and processing speeds increased substantially. Software written specifically for point cloud management helped to overcome some of the processing limitations. However, as scanners evolved their data evolved as well, producing point cloud files recording millions of points and reaching multiple gigabytes in size.

Today's computers have a level of processing speed unimaginable ten years ago. Along with considerable improvements in graphics cards, computers can now routinely handle large graphic files. Many of the computer's improvements have been due to the demand for high-speed gaming computers, capable of handling large, constantly changing graphic files. This has been an asset for those working with point cloud data since processing and using point clouds involves constant regeneration of the screen image.

Data acquisition speeds have increased and point clouds captured for large projects are routinely hundreds of millions of points, or even more. A point cloud file for an entire factory could be several terabytes worth of data, the point cloud of a smaller project could be hundreds of megabytes. Aside from state-of-the-art technology within a computer, to process point cloud data one needs a powerful software specifically designed to handle point clouds and structured mesh files. Often multiple software are required to assist one in the various stages of point cloud preprocessing, registration, and alignment.

There are a few useful functions needed when processing point cloud data. They are:

- Coordinate System: defining a coordinate system or systems.
- Scaling: transforming the dimensions of the point cloud.

3D Scanning for Advanced Manufacturing, Design, and Construction, First Edition.
Gary C. Confalone, John Smits, and Thomas Kinnare
© 2023 John Wiley & Sons, Inc. Published 2023 by John Wiley & Sons, Inc.

- Mapping: aligning point clouds in proper relation to one another
- Uncertainty: determining allowable levels of precision and accuracy

Many of these functions can be performed inside measurement or registration software. They can also be performed in point cloud editors, reverse engineering software, or inspection software. But not all software can provide these functions. It's important to know the capabilities of both your software and the end user's software, as well as file formats required, so that you can provide a digital deliverable that meets their needs.

5.2 COORDINATE SYSTEMS

A coordinate system allows data to be given a numeric description; in surveying it allows data to be aligned to a local geo-reference or to an established USGS datum [United States Geological Survey point]. It can also be used to align a point cloud to features in the real world.

An example would be an architectural scan where control points surveyed during the scan are matched to previously measured control points on-site. After aligning the two sets of points, moving and rotating the scan data so that the control points match, the resulting scan data is matched to the geodetic control system previously established on-site. When using the scan data, aside from knowing the four compass directions, one can describe the position of any point on the site, the corners of a building, or the grade elevations in relation to a fixed datum height. A surveyor could survey and scan the site of an adjacent building, align his scan to the same state plane coordinate system via control points, and be able to ascertain the closest points between the two buildings.

Coordinate systems can also be defined by constructing features through regions of points in the point cloud. This approach is often used in consumer product reverse engineering scenarios. Regions of the cloud that can be modeled as geometric primitive shapes like planes, cylinders, or spheres are selected, and these features are fit to the cloud. In some software planes of symmetry can be solved as a feature. The features can then be used to define the coordinate system. The point cloud data can then be exported in that coordinate system. This can make reverse engineering much easier, and presents the point cloud or model to the end user in a more useful manner than if it were randomly oriented in space.

Another method to achieve a coordinate system is by best-fitting regions of point cloud data to an existing CAD model. This approach is often used in parts inspection. One advantage to this approach is that the CAD surfaces don't need to be geometric primitives; they could be complex curved surfaces like a car body panel, a boat hull, or a turbine blade.

If the coordinate system is not defined by an existing model, you may need to reference the blueprint or drawing in order to define the proper coordinate system for the subject under investigation. For example, aircraft, ships, and vehicles all have coordinate systems to help describe the vessel's shape and size.

Before explaining the industry practices for coordinate systems, it is best to get an overall view of the three basic styles that you may encounter. These include the cartesian coordinate system, the cylindrical coordinate system, and the spherical coordinate system.

5.2.1 Cartesian Coordinate System

The cartesian coordinate system is by far the most frequently seen of all the frames in the world today. This system is the default for many of the CAD packages engineers and architects use to design parts. It is hopeful that the part will be designed with manufacturing and inspection in mind in order to use the benefits of this coordinate systems.

The cartesian coordinate system consists of three coordinate axes mutually at right angles to one another (Figure 5.1). These axes are identified as x,y,z axis and are customarily identified in the right-hand coordinate system. What this means is that when rotating the x-axis into the y-axis, a right-handed screw would project into the

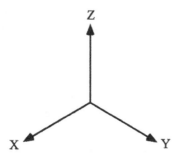

Figure 5.1 Cartesian coordinate system consists of three coordinate axis x,y,z

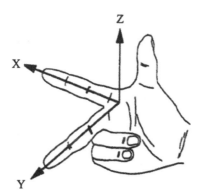

Figure 5.2 Right-hand coordinate system

z-axis denoting the rotational axis of a CNC machine. This reference system is taught in the engineering and machine tool trades where the z-axis is referencing the axis of rotation (vertical axis) referred to as the axial plane.

Figure 5.2 shows a right-hand coordinate system. A simple way to remember this is by using the right hand to identify each axis where the pointer finger is the x-axis, the middle finger is the y-axis, and the thumb is therefore the z-axis. In a left-hand coordinate system, the same would hold true. Note that the direction each finger indicates would be the positive value direction for this axis and the origin of the coordinate system is the intersection of all three axis.

5.2.2   Cylindrical Coordinate System

With cylindrical shapes and parts, it is often simplest to identify a position by calling out the distance from the centerline and an angle off the reference axis. Of course, you could use the Cartesian coordinate system and trigonometry to identify features and locations on this cylinder, but if the part is axisymmetric, it is best to use the cylindrical coordinate system and identify a centerline and a reference plane, where the intersection of these two features is your origin. In Figure 5.3 the z-axis is the centerline axis and the x-y plane is your reference plane. The radial distance from the centerline is identified as (r), and the angle off of the reference axis (x-axis) is labeled (θ). These two features (r, θ) are also known as polar coordinates. Lastly the height off the reference plane (x-y plane) as illustrated in Figure 5.3 is identified here as Z. Note, the origin of the cylindrical coordinate system is typically identified as a primary reference plane in the part under investigation.

The Cartesian coordinates (x,y,z) and the cylindrical coordinates (r,θ,z) of a point are related as follows:

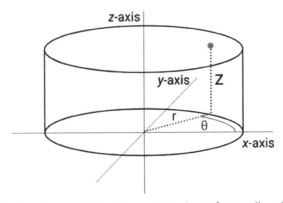

Figure 5.3 Cylindrical coordinate system consists of r = radius, θ = angle, z = height

- $x = r\,(\cos\theta)$

- $y = r\,(\sin\theta)$

- $z = z$

5.2.3 Spherical Coordinate System

A little more challenging to grasp is the spherical coordinate system. This coordinate system may be difficult to understand at first, but once faced with a spherical object which needs key features on the surface identified, it makes perfect sense. Spherical coordinates determine the position of a point in 3D space based on the radial distance "r" from the origin and two angles "θ" and "φ."

To dissect this further, it may be easiest to think of it this way. Calculate the relationship between a Cartesian Coordinate (x,y,z) of the point identified in Figure 5.4 as point "p" using trigonometry. Its spherical coordinates are (r, θ, φ). If you look at it in two dimensions, the length of the leg of the right triangle in the y-z plane is the distance from p to the y-axis, which is r(sin θ). The length of the leg of the right triangle projected onto the y-axis r(cos θ) is the distance of p from the z-axis. The same scenario holds true for the x-y plane trigonometry.

The Cartesian coordinates (x,y,z) and the spherical coordinates (r, θ, φ) of a point are related as follows:

- $x = r\,\sin\varphi\,\cos\theta$

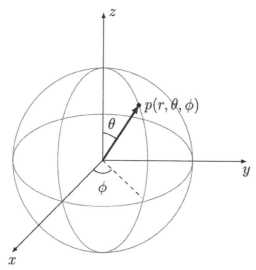

Figure 5.4 Spherical coordinate system consists of r = radius, θ = angle, φ = angle

- $y = r \sin\varphi \sin\theta$

- $z = r \cos\varphi$

5.2.4 Aircraft Coordinate System

When applying these coordinate systems to real-world applications the reasoning behind them becomes apparent. Let's examine one of the most popular industries where precision metrology is commonplace. In aerospace design engineering, manufacturing, and quality control, accurate representation of the aircraft by defining its coordinate system is essential. The aircraft coordinate system is not much different from other industries in that it uses a Cartesian coordinate system although this industry has adopted a name convention to make things clearly understood. The x-axis is referred to as the Station Line on and aircraft and the z-axis is referred to as the Waterline. Where the z origin is typically below the landing gear with a positive value upward, and the x origin is located a distance from the nose of the aircraft with the x positive value increasing as you head aft. The y-axis is at a right angle to the centerline of the aircraft, the butt-line as it is referred. To avoid any confusion from one side to the other, it has a left and right side referred to as the left butt-line (LBL) and the right butt-line (RBL). This is illustrated as a helicopter coordinate system in Figure 5.5.

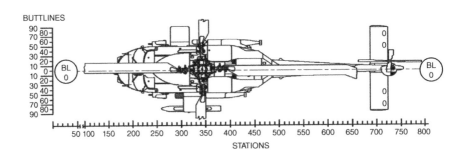

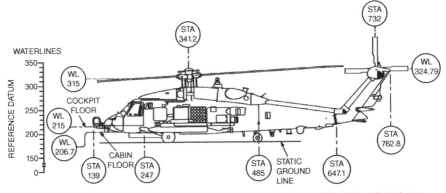

Figure 5.5 Aircraft coordinate system consists of X = Station Line (STA), Y = Buttock Line (BL), Z = Water Line (WL)

Source: Image courtesy of Sikorsky Aircraft Corporation.

5.2.5 Surveying Coordinate System

Surveyors use a global and local Cartesian coordinate system depending on the application. In a global system, or geocentric terrestrial system, the origin is located at the center of the Earth and the primary axis of the system is aligned with the Earth's rotational axis. The secondary axis is based on a conventional definition of the zero meridian. Lastly, the tertiary axis is at a right angle to the primary and secondary axis (Figure 5.6).

When surveyors use a local or topocentric coordinate system, an origin is defined by a chosen point on the surface of a plane. Often this is a marker in the ground called a bench mark. The y-axis of a topocentric system is pointing in the North Direction, and the z-axis is pointing upward most often but not necessarily normal to the plane defined by gravity. This leaves the x-axis pointing in the approximate direction of East. Surveyors and architects will often refer to these as Easting, Northing, and Elevation instead of X, Y, Z respectively. Most often surveyors are collecting data in 2D polar coordinates. The instruments they use are typically total stations and theodolites to collect point and azimuth angles from a reference line. Scanners frequently use the coordinate points defined through survey methods as a means to tie their collected data into a recognizable coordinate system.

5.2.6 Degrees of Freedom (DOF)

We often hear the term 3-DOF or 6-DOF while working with coordinate frames. DOF is simply the degrees-of-freedom, thus 3 DOF means 3 degrees of freedom and 6 DOF means six degrees of freedom. A 3 DOF system allows the translational freedom of movement along the three-coordinate axis x,y,z. When a system has 6 DOF, it adds a rotational component to the three axis (Figure 5.7). Each of these rotational movements are identified as follows:

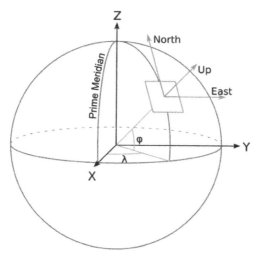

Figure 5.6 Geocentric terrestrial coordinate system consists of X = East, Y = North, Z = Upward

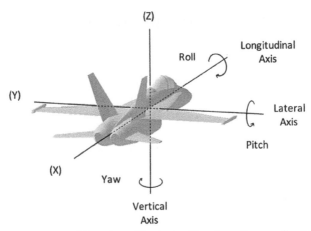

Figure 5.7 Six degrees of freedom illustrated in aircraft coordinates
Source: CHRobotics 2012.

1. Roll is the rotation or "roll" around the longitudinal axis (x-axis). For an aircraft this is the rotation around the centerline of the fuselage.

2. Pitch is the rotation about the lateral axis (y-axis), or the nose up and nose down rotation.

3. Yaw or azimuth is the rotation of the aircraft around the vertical axis (z-axis). This can be perceived as the directional vector or orientation of the aircraft.

For naval vessels the six degrees of freedom are similarly referred to as roll, pitch, and yaw for rotational motion, but the translation is referred to as surge, sway, and heave corresponding to x,y,z respectively (Figure 5.8).

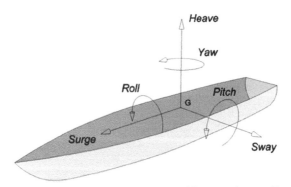

Figure 5.8 Six degrees of freedom illustrated in naval coordinates

5.3 SCALING

Scaling the point cloud means changing its size. This is often not necessary if the points were produced by a scanner's measurement system that directly determines size during the scanning process. Scaling is essential for measurement systems like photogrammetry that don't directly determine size during measurement. In this case the most common method of scaling is to include an artifact or artifacts of a known size or distance apart in the scan or photo. During post-processing the artifact or artifacts are identified and the size or distance is input into the software, thus setting the entire group of data to a fixed scale.

Another reason for scaling can be found in high precision inspection applications. Objects tend to grow when heated and shrink when cooled. This means if the part is measured at a temperature that is not the *reference temperature* (20°C for most mechanical parts), the part may measure too large or too small, and thus outside specified tolerance. The data must be scaled to compensate for the tolerance variation caused by errant temperature.

In some cases, it can be very useful to scale the point clouds to account for changes in units. In point cloud file formats that do not communicate their units, it is often helpful to use scaling to convert data into the desired unit format. For example, some long-range scanners record and export their data in meters. If the intended final use is to be in inches, then we would want to scale (convert) the point cloud in the post-processing software to the proper units.

5.4 MAPPING TECHNIQUES

Mapping, aligning, and registering are all terms used to describe putting scans together so that features common to the scans coincide. It can also refer to aligning a point cloud with known features that were established from a previous survey, inspection, or CAD model.

In modern practice, cloud to cloud fitting is the most prevalent method of aligning scans. The software takes two or more scans, then moves and rotates them around each other using complex algorithms until it achieves the best possible fit. This is often achieved using a least-squares iteration to minimize point-to-point deviations.

Though scanners are extremely accurate, slight errors will be detected in certain regions or angles in the field of view. This is inherent in both rotational scanners and fixed scanning systems. In rotational systems errors will typically be seen in the rotational encoders. This is primarily caused by axial runout. In stationary systems errors will occur from optical aberrations in the lenses. These are normally found at the extreme edges of the field of view. Compensation routines which are performed prior to the scanning process will accommodate any large errors in the system, but there will always be slight errors that are permissible by design. The manufacturer specifications will list a specification for the maximum permissible error (MPE). The MPE is used during the calibration process to allow the laboratory to get the system within the manufacturers accuracy specifications. These specifications are what is necessary to define the capability or overall accuracy of your system.

When performing cloud-to-cloud fitting, the system errors will not be noticed unless you are familiar with the specifications of the part being measured, or if you are performing a closed-loop scan. In the latter case, you would see a mismatch in data when joining the last set of point cloud data to the first set collected. In Figure 5.9 the laser scanner is positioned in seven different locations around the automobile to collect all the vehicles features. If a cloud-to-cloud fit procedure is used, slight errors may not be noticed from one scan to the next until the complete closed-loop scan is formalized. At this point the engineer analyzing the data may notice a step between the point-cloud set of the first scan, and the point-cloud set of the final scan.

To better understand how this type of error can impact your measurements consider the following example.

Here is a situation that occurred in the earlier days of scanning: Cloud fitting resulted in a slow curling of multiple combined point- clouds. The subject being scanned was a 4-feet granite surface plate which was known to be flat to 0.001 inch. After the cloud-to-cloud fit was performed, the overall transition from one cloud to the next was extremely smooth, yet the system indicated that the surface had a 0.020-inch bow in it. This bow was due to the stack up of system error from one scan to the next.

Modern software does a much better job than those early versions, and in many cases today, cloud-to-cloud fits are the best way to register scans into a single file.

To counter the above condition, engineers and software developers derived a very accurate way to combine or "register" scans using common points. Depending on the

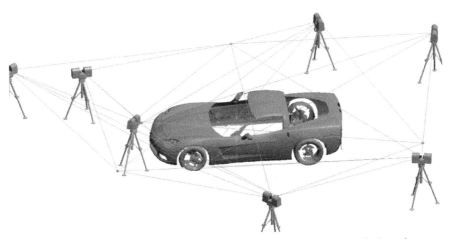

Figure 5.9 Multiple scan positions insure adequate coverage of all surface areas to be documented
Source: Image courtesy of Hexagon Manufacturing Intelligence – Spatial Analyzer.

industry and the region, common points are also called, control points, benchmarks, fiducials, or monuments, among a long list of names. When using common points to register scans, they should be located within the envelope of the survey area and measured with a higher accuracy instrument such as a laser tracker or a total station. The high accuracy common points will provide a reliable network for the scanners to be positioned and simplify data analysis back in the office.

The combination of a laser tracker and a laser scanner are illustrated in Figure 5.10 where common points are located at ground level in order for the scanner to move about within the predefined coordinate frame.

The process of using common points is best described as follows:

- A network is laid out which will allow for line of sight within the survey or part envelope. It is often best practice to spread out common points on the part being measured. Straight line or coplanar patterns should be avoided. If scanning a building, it is most likely points will need to be located on the walls and floor. The key component is that the common points are attached to a stable surface, and they must not be tampered with until the survey is complete.

- Next, value the network using a laser tracker system or a system with a high degree of accuracy.

- You are now ready to scan the subject or area.

Once all the scan data is collected, merging the files is completed by having corresponding common points (which were valued by the laser tracker), identified in each scan data set that is going to be merged. The overall network of common points provides an accurate reference frame for each scan to be placed accordingly. The common points from the scans will be best fit to the common points of the network frame with extreme precision.

This method can help prevent an accumulation of error from one end of a job to another, but the clouds transitions will not be quite as smooth as the cloud-to-cloud fit

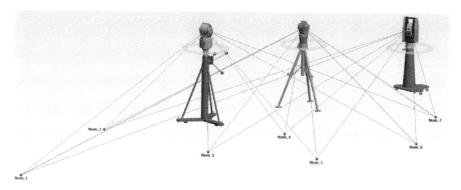

Figure 5.10 Through the use of common points, a network can be created to document the laser scanner data with extreme accuracy
Source: Image courtesy of Hexagon Manufacturing Intelligence – Spatial Analyzer.

solution. You may even see "double walls" in regions of the combined point cloud where two or more scans overlap. This overlap is due to the scanning instruments accuracy which would otherwise go unnoticed when performing cloud-to-cloud fitting. The difference being that the scanner error is spread evenly throughout the network in this common point process, whereas in the cloud-to-cloud process the error is only noticeable between the first and last scan set merge and it can often be somewhat dramatic.

In tight tolerance manufacturing it is best to use the common point process of fitting scans for both large- and small-scale jobs. But, in the AEC industry where accuracy may not be critical, the cloud-to-cloud fit process is widely accepted and far more practical.

Aside from accuracy, one other advantage of common point fitting is that it is easier to diagnose cloud registration issues if they may exist, or more simply stated, how well the clouds are joining together. In cloud fitting software you can view a statistical report showing how well the adjoined clouds went together, but you will see no indication of any trends in the data. This differs from common point solutions where one is able to calculate the magnitude and direction of the deviation of any common point in the high accuracy framework.

Worth mentioning, a third method of registering clouds is to manually identify corresponding regions of the clouds by windowing portions of the cloud or by fitting features to regions. This method requires care in selecting the regions to fit. Ideally you want to fit large, easily identifiable regions so that you don't accidentally choose regions that don't overlap well.

In the long-range scanning world, there are some interesting developments. Many scanners are now equipped with GPS receivers. When you take a scan, the instrument

Figure 5.11 Advances in software and hardware have made reverse engineering tasks such as the modeling of this aircraft much more efficient than was achieved only a decade ago
Source: Photo courtesy of Oqton.

records its location and orientation. Other scanners have cameras or other sensors that track their motion between scans. Either of these techniques can provide a rough registration of the scans to each other, and in the case of GPS, they can be registered to a known coordinate system. Unfortunately, these systems are not currently accurate enough for a final registration, but they do provide a good starting point to save a lot of time during registration in post-processing.

Aligning point clouds to design coordinate systems or CAD models will employ a similar technique as that seen in the registration of scan clouds to one another. By using a CAD model with a predefined built-in coordinate system, as illustrated in Figure 5.11, the aircraft scan data can be best-fit to the model surface features. Once this process is completed within the software, each point within your scan cloud will be valued to correspond to its location in the aircraft coordinate system.

5.5 CAD TO PART COMPARISON

As stated previously, one of the most common and useful analyses that you can do with a point cloud is to compare measurements from an item with the ideal shape as defined by the designed CAD model. When the closest point on the CAD surface to the measured point is identified, then it is relatively simple to calculate the deviation of the measured point to the CAD surface.

For example, take the coordinates of the measured point, $x_m, y_m, z_m,$ and the coordinates of the closest CAD point, $x_c, y_c, z_c.$

Then you subtract corresponding coordinates to find the components of the vector between the two points.

$$dx = (x_{m-}x_c), dy = (y_{m-}y_c),\ dz = (z_{m-}z_c)$$

Now you can find the distance, or the deviation between the measured and the CAD points by using the root sum square equation. Root Sum Square (RSS) is a generic term for any equation of this form. (If we were considering distance in a single 2D plane, the RSS is called the Pythagorean Theorem.) To calculate the RSS square of each of the components above, add the squares together, and then take the square root of the sum as shown below:

$$Distance\,(deviation) = \sqrt{(dx^2 +\ dy^2 +\ dz^2)}$$

Once the deviations have been determined for all the measured points the software can generate a slew of data describing the deviations as a group. Some of the simpler and more useful values are the maximum (max) and minimum (min) deviations. These are simply the deviations of the measured points that are farthest above and farthest below the nominal surface respectively.

Be careful, however, the minimum deviation should be considered as a signed value where negative (−) values indicate the measured point is below the nominal surface. Some software doesn't consider the measurements as signed values and reports the minimum as the deviation of the point closest to the nominal surface.

Most software can also calculate an average deviation. This is another useful number for determining the quality of a surface. Some software will calculate average of the positive deviations and the average of the negative deviations. These averages can help give you a feel for the spread, or distribution of the deviations.

5.6 ROOT MEAN SQUARE AND STANDARD DEVIATION

Two other values can be calculated that are often good indicators of the spread of a deviation. They are *root mean square* and *standard deviation*. Both of these provide more meaningful results when they're applied to large numbers of points, so they are well suited for scanning applications.

The method to calculate the root mean square of the deviations is to:

1. Square each deviation

2. Add all the squares of the deviations together

3. Divide that sum by the number of deviations

4. Take the square root of the resulting quotient

Mathematically the process is represented by this equation.

$$d_{rms} = \sqrt{\frac{\sum_{i=1}^{N} d_i^2}{N}}$$

Where,

d_{rms} = the root mean square of all the deviations
i = an index or counter for the deviations
N = total number of deviations
d_i = the ith deviation
Standard deviation, the darling of the statisticians, has a very similar calculation,

$$\sigma = \sqrt{\frac{\sum_{i=1}^{N} (d_i - d_{ave})^2}{(N-1)}}$$

Where the variables are the same as for RMS above with the addition of

σ = the standard deviation of the deviations
d_{ave} = the average of the deviations

The steps in calculating the standard deviation are:

1. Take the difference between each deviation and the average of all the deviations
2. Square each of those differences
3. Add all the squares of the differences together
4. Divide that sum by one less than the number of deviations
5. Take the square root of the resulting quotient

5.7 THE METHOD OF LEAST SQUARES

Least squares is an important means of fitting measured data to idealized shapes or point sets, or vice versa. The general idea is to minimize the sum of the squares of the deviations of the individual data points. It has several desirable properties:

1. Easy to program
2. Stable solution
3. Results close to a minimized deviation approach, especially for "good" data
4. Well suited for projects concerned with RMS or standard deviation relative to a CAD surface model

Of course, nothing is perfect. Least squares solutions may not be exactly what you are looking for as a result. Often times we are interested in minimizing extreme deviation. This is certainly the case in over 95% of inspections where everyone, the machinist, the engineer, the project manager, and the inspector would really prefer that all deviations fall within the design tolerance. Thus it is important to know that least squares solutions do not minimize the extreme deviation, instead it minimizes the sum of the squares of the deviations.

5.7.1 Other Geometric Fits

Sometimes least squares fits do not provide the answer that you really need. For example, if you measure a hole as a cylinder and use a least squares fit to determine the size of the diameter, the fit gives you an average value. Some of the measured points lie inside the diameter, some lie outside. If you try to run a pin the size of your least squares diameter into the hole it won't fit because it will be obstructed by the material represented by points inside the least squares diameter. If you want to know the size of the largest pin that will fit in the hole, you must use a maximum inscribed fit.

Thinking about this kind of fit geometrically, we might use the following process to find the maximum inscribed cylinder.

1. Start by calculating a least squares fit for the hole.
2. Calculate a cylinder with a small diameter using the centerline of the least squares cylinder.

3. Iteratively expand the cylinder until it contacts a measured point.

4. Continue to iteratively expand the cylinder but allow the cylinder centerline to move proportionately away from the point of contact as it expands so that the first contact point remains a single point.

5. Cease expanding the cylinder when 4–6 points of contact have been established. (Four points if the top and bottom of the cylinder encounter diametrically opposed points. Six points if the top and bottom expand to three points of contact each.)

A minimum circumscribed fit operates similarly, but its goal is to find the smallest diameter ring that will accommodate a measured shaft.

Another useful type of fit modifies the least squares method by constraining a parameter. Fixing the radius of a circle or cylinder to a desired value during the fit is a very common technique. It can be used to try to obtain a better value of the center of a misshapen round feature, or to remove difficulties inherent in solving the radius of an arc that subtends a small angle, i.e., trying to define a circle using points that cover ¼ of the circumference.

Finally, it is often desirable to find a solution that minimizes the range of deviations. This is the condition called for when evaluating Geometric Dimensioning and Tolerancing (GD&T).

The process would look something like this:

1. Solve the least squares fit for the measured data.

2. Find the points of maximum and minimum deviation.

3. Calculate the effect of small rotations and translations that reduce the difference between the max and min deviations.

4. Reorient the feature to a new candidate location and repeat the process.

5. Select the orientation and position that minimize the difference between the max and min deviations.

5.7.2   3D Best-Fitting

3D best-fitting uses the method of least squares to move a set of points into alignment with another set of points or surfaces. A common use of this is to orient the different stations or locations of an instrument when it is relocated.

This is achieved when "control points" (datum points, benchmarks, etc.) are measured in the initial position, or measurement station. All measurements are now taken in the coordinate system of the instrument in its first station.

The instrument is then relocated to its new station and the control points are measured again. These points are described only in the current coordinate system of the instrument. If coordinates of the points from the two stations were compared there would be large differences between the stations.

Then a least squares fit is used to calculate the transform of the translations and rotations that will most closely align the control points from the second station to those from the first station. The transform is applied to the second station instrument and all its measurements. Now measurements made in the second station can be reported as if they had been made in the first station. A uniform frame of reference has thus been established.

A comparatively small number of points, seven to a few dozen, are typically used in control point fits. Note, no less than three points can be used to complete this transform.

The exact same technique can be used to align an instrument with nominal points, previously measured known points, design points, or the centers of tooling balls.

5.7.3 Weighted Fitting

Weighted fitting is another powerful technique built around least squares. It can be used in the same ways as described above, but in weighted fitting predefined parameter constraints are placed on the solution. This can take at least three forms.

1. In the first form individual points are weighted, usually on a scale of 0 to 1. Depending on the software, weighting can be applied individually to control translation in the x-, y-, z-axes or to control rotation around these axes (i, j, k). Typically used only in control point applications, not for clouds. Points or components set to low deviations will have less effect on the overall fit. An example of where this might be useful is if you have four control points but only three of them define the primary (XY) plane of orientation. The Z-axis of the fourth point would be set to 0. Then it could affect the XY location of the origin but would not pull the primary plane out of alignment.

2. Overall fits are constrained in a number of degrees of freedom. This can be used with cloud to surface, control points, feature-based, or coordinate system-based fits. It's typically used when you have one aspect of the alignment suitably established but wish to refine other aspects. For example, a shaft centerline has been aligned with the x-axis of the coordinate system, but it still needs to be clocked around the axis and set to the correct translation along the axis. In this instance you would use a best fit instructing the software not to translate along the y- and z-axes, or rotate around these axis. This weighted fit condition would then leave the fit free to slide along the x-axis and rotate around it.

3. Weights established by instrument model is a very powerful technique. It is sometimes called bundling. This method uses a model of the instrument uncertainty to weight the parameters of the instrument which are more accurate. It is very computation intensive and therefore used only with control points, but it can reduce the uncertainty of a network of control points significantly. For example, a laser tracker collects points by measuring two angles (azimuth and elevation) and a distance (x-axis) from the laser head. It is far more accurate in the x-axis which is

measured using time-of-flight technology or interferometry. Again, this axis is the distance away from the laser tracker head. Software designed to compensate for the strength of the distance parameter over the azimuth and elevation angle calculations are then taken into consideration.

New River Kinematics provides a highly sophisticated software package called SpatialAnalyzer (SA) that will deliver this type of high accuracy solution for network alignments, particularly when various instruments come into play to collect one set of data. USMN (Unified Spatial Metrology Network) is a bundling technique that optimizes instrument and target positions by intelligently using measurement uncertainty weighting, taking into consideration the accuracy of the instrument, the distance between the target and the instrument, and a number of other environmental and operational parameters.

5.8 WHAT IS UNCERTAINTY?

Estimating the uncertainty of a measurement is often an overlooked aspect of 3D scanning. It is an effort to estimate the accuracy level of the measurements taken during the scanning process. When taking measurements, we assume there is an exact value or representation of the subject being measured, but we never know what this exact value will be. We can apply different methods and equipment to come up with a range of measurement results that can be used to report our best guess or estimate of the exact value. This range of values is defined as the measurement uncertainty.

In many applications defining the uncertainty is not required, or it is not always essential, for the project to have a successful outcome. In these circumstances, the accuracy requirements may be extremely low, and the uncertainty estimation may be so complex and time consuming it is unnecessary and becomes cost prohibitive. For circumstances requiring a high degree of accuracy and measurement reliability, uncertainty calculations are extremely important and an essential part of quality control.

There are two main reasons for determining measurement uncertainty.

One of the primary reasons for determining uncertainty is to define the capabilities of your equipment, operator, and the scanning environment prior to collecting the measurement data. This calculation may help to determine which instrument is best suited for the application based on measurement tolerance requirements and an environment that may not be under tight control. Environmental factors that can hamper measurement accuracy, particularly with laser scanners, include temperature, pressure, and humidity.

Secondly, uncertainty can be calculated after the data collection or scanning has occurred. This uncertainty calculation is based on the actual environmental conditions recorded over the course of the job, and it will include the capability or accuracy of the measurement system. This data will be used as the first link of measurement traceability, ensuring that the final measurement results are taking into consideration all of the factors which may negatively contribute to the integrity of the data.

A statement of uncertainty is vital to communicating the quality of a data set. Such a statement can help analysts determine whether to trust one set of data more than another; that is important if two sets of measurements conflict. Such a statement can help an engineer determine whether a particular scan is sufficiently accurate to really determine the conformance of a scanned part to its design by asking one simple question, "Is the scan accuracy of sufficiently smaller magnitude than the tolerance?"

Determining uncertainty can be an involved process. There can be a number of factors influencing the level of uncertainty found in a project. Consider at least the following when determining an estimation of uncertainty:

- Uncertainty of the instrument as determined by the instrument's specification.

- The effects of environment – temperature, humidity, vibration, lighting, dust.

- Stability, both mechanical with respect to setup, and environmental, during measurement.

- Operator influence.

- Range effects. (Does the uncertainty of measurement change with distance from the instrument to the surface being scanned.)

- Other extraordinary factors.

Because uncertainty calculations can be complicated and depend heavily on the conditions at a particular site, most software does not have a function to estimate it. A simple statement of instrument accuracy as specified by the manufacturer in tandem with environmental conditions at the time of measurement may be sufficient.

Below is a simple example of one method used to calculate uncertainty for the length measurement of a baseball bat. The uncertainty scale is exaggerated for clarity. The first step would be to take a series of measurements of the baseball bat.

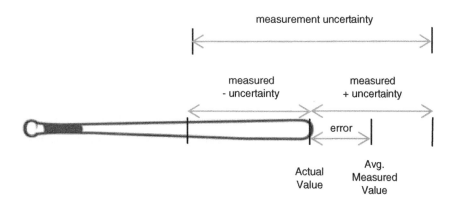

Taking five measurements of the length of the bat we get:

$$X1 = 32.32''$$
$$X2 = 32.34''$$
$$X3 = 32.43''$$
$$X4 = 32.36''$$
$$X5 = 32.38''$$

The average measured value of the bat would be equal to

$$Average\ Measured = \frac{X1 + X2 + X3 + X4 + X5}{5}$$

Average Measured = 32.37″ (always round to the same significant figures as the measurands).

Next find the variance of the measured values by first determining the difference of each value from the average measured value. The variance is the average of the sum of the squares of these differences calculated as follows:

$$V1 = X1 - AVG\ MEASURED = 32.32'' - 32.37'' = -0.05''$$
$$V2 = X2 - AVG\ MEASURED = 32.34'' - 32.37'' = -0.03''$$
$$V3 = X3 - AVG\ MEASURED = 32.43'' - 32.37'' = 0.06''$$
$$V4 = X4 - AVG\ MEASURED = 32.36'' - 32.37'' = -0.01''$$
$$V5 = X5 - AVG\ MEASURED = 32.38'' - 32.37'' = 0.01''$$

$$Average\ of\ Squares = \frac{V1^2 + V2^2 + V3^2 + V4^2 + V5^2}{5}$$

$$Variance = \frac{0.072}{5} = .0014$$

Next calculate the standard deviation from the variance by taking the square root

$$Standard\ Deviation = \sqrt{(Variance)}$$

$$Standard\ Deviation = .038$$

The final measurement then is equal to 32.37 ± 0.04.

These calculations are typically performed within the algorithms of the software and in scanning will consist of hundreds of points. Point averaging is automatically done with the results reported on geometric features such as planes, circles, cylinders, and spheres. When performing a scan of a modeled surface for point cloud to CAD comparison, the deviations are in relation to the designed surface and depending on the programmed constraints will reflect the average deviation or best-fit solution to the surface.

5.9 CONTRIBUTING FACTORS TO UNCERTAINTY

There are many contributing factors to measurement uncertainty, some of which you will face on every job. Below are four of these parameters with a brief description outlining the impact and the steps you can take to mitigate the impact on your measurement uncertainty.

5.9.1 System Accuracy

System accuracy will impact every measurement that you perform in one way or another. The manufacturer will specify the overall system accuracy, but it is best to know what the major contributors are during the operation of each instrument. When using articulating arms, each articulating joint contributes some error to your final result. In order to reduce this error, one must take care to minimize the rotation of these joints to the best of their ability and still capture the data required. With all instruments it is often best practice to relocate or reposition the instrument using common points to lock back into the original coordinate frame rather than stress the device to its operational extremes. Laser tracker systems are characterized by having superior distance measurement capability within a specified range. The time-of-flight units and in particular the interferometer-based trackers are capable of achieving submicron accuracy. It is therefore best to set these units up to measure in a position that favors this capability. For instance, when measuring an aircraft wing, you would want to position the laser tracker at the end of the wing prior to scanning. This would allow the operator to collect most of the data using the more accurate parameter of the laser tracker which is the length axis. This is counterintuitive to the old days of operating standard optics where the technician would benefit by placing the theodolite in the middle of the wing to increase the apex angle and in turn increase the accuracy.

5.9.2 Environmental Effects

Although the distance resolution of a laser tracker is more accurate than the angular component (azimuth and elevation), there are environmental conditions that will degrade this accuracy. With any laser one must be aware of the impact the measuring environment can have on your results. This is particularly true for terrestrial laser scanners and laser trackers which are accustomed to taking measurements more than a meter from the source.

5.9.2.1   Environmental Parameters and Wavelength Distortion

Most of the high-end systems today contain weather stations which monitor and scale results based on the environment temperature, barometric pressure, and humidity. The impact these environmental parameters will have on a wavelength-based measurement system are as follows:

Temperature: 1 ppm / 1°C

Pressure: 1 ppm / 2.5 mm Hg

Humidity: 1 ppm / 85% change

Be aware, however, the weather station only measures the environment around the laser source unless one is employing thermocouples to monitor the entire measurement area.

Temperature gradients will distort the laser beam much like the effects seen when looking at an object submerged in water (Figure 5.12). In this instance, the laser beam is traveling through a median of varying density which will introduce refraction. This is often hard to avoid with large-scale scanning projects, but a good metrologist will understand the uncertainty and the impact it could have on the end results. For high accuracy scanning it is best to avoid any significant temperature gradient by keeping the system and the subject under study relatively close to the laser source. Temperature gradients of 2° C across a measurement plane can introduce an uncertainty error of 0.0254 mm at a range of only 5 meters. Refractions are commonly seen on the highway on a hot summer day. Sometimes called heat haze, an object will appear distorted as the heat from the asphalt distorts the view of a distant object. The refraction or distortion of light in this instance demonstrates the similar impact it would have on a laser beam and thus your measurement results.

5.9.2.2   Material Property Effects on Measurement Accuracy

All materials have a coefficient of thermal expansion and contraction, and a coefficient of thermal conductivity. The coefficient of thermal expansion dictates how much the material will change with temperature, and the coefficient of thermal conductivity

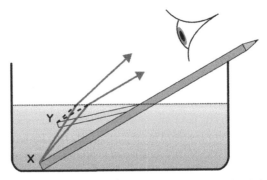

Figure 5.12 Refraction at the water/air interface and optical density differences account for the "bent pencil" illusion

determines how fast the material will respond to changes in temperature. Knowing both parameters can be useful in determining whether the measurement data is reliable.

The coefficient of thermal expansion (CTE) denoted by the Greek letter $\propto$ (alpha) relates how much a material's linear dimensions will expand for each degree of temperature change (Figure 5.13). The CTE is in other words, a fractional change in the length of a bar of the material per degree of temperature change as given by the formula below:

$$\propto = \frac{\Delta L}{L \times \Delta T}$$

where,

ΔL = the change in length of material in the direction being measured
L = overall length of material in the direction being measured
ΔT = the change in temperature over which ΔL is measured

Material	CTE in 10^{-6} /°C
Steel	11.39
Aluminum	22.39
Titanium	8.82
Invar	1.26

Figure 5.13 Coefficient of thermal expansion of some common materials used in manufacturing

Thermal expansion or contraction is not the only factor which can distort your scanning subject. Distortion from stress or load can often cause parts to change significantly in shape and contour over the course of a measurement job. Exposure to sunlight or a temperature source (heater, overhead door, etc.) will introduce challenging circumstances in uncontrolled environments.

If you spend enough time in an inspection department, you will hear folks say that the part must "soak" prior to dimensional measurement. What they are referring to is the stabilization of a part to ambient temperature. Depending on the material, density, and size of the part, it is typical practice to allow the part to reach design temperature prior to measurement. If you learn nothing else from this book, the most important take away is that you know all designs and blueprints are characterizing the part, building, or assembly at the temperature of 68 degrees Fahrenheit (20 degrees Celsius). Though most software today will request the operator enter the material being measured, it is good practice to understand the various material properties and how they are effected by temperature.

When reverse engineering a sheet metal or composite panel it is often best to perform this scan in place or onboard the vessel it is mounted. Unless you can guarantee its shape conforms to the original condition, these parts will often deflect or spring when removed and cause the scanner to capture an inaccurate representation of the parts design configuration.

5.10 TYPICAL POST-PROCESSING WORKFLOW

The first step in post-processing is often to clean up point cloud data by removing scan points that are not part of the desired scan region. With long-range scanners, this could be points captured on adjacent buildings or structures. With short-range scanners the points could be part of the table top or fixture that supported the scanned object. These extraneous points are often quickly and easily identified and removed manually.

A common second step in early post-processing workflows was to reduce "noise" in the point cloud by employing a statistical filter that removed outliers, points more than a certain distance from the average surface of the scan. Then the part would be decimated, also known as sampled. That is, the number of data points in the cloud would be reduced using one of several strategies.

In modern practice the second step is usually to sample the point cloud in lieu of noise reduction. There are two reasons for this: First, contemporary scanning equipment collects so much data that we often don't need all of it. Sampling reduces the total number of points, thus speeding up subsequent operations. Second, data acquisition in today's scanners often includes some form of noise reduction or other kind of quality

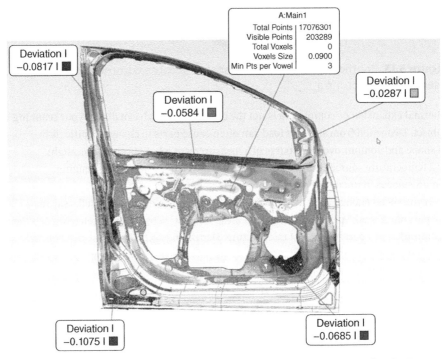

Figure 5.14 Random point cloud sampling can be cumbersome for CAD modeling, but it is successfully used for CAD to part comparisons when performing inspection tasks
Source: Image courtesy of Hexagon Manufacturing Intelligence – Spatial Analyzer.

check during the actual scan process. Therefore, the raw point cloud data may need little or no noise reduction.

There are three main filtering strategies employed in sampling. Random sampling, uniform or grid sampling, and curvature-based sampling.

Random sampling is the fastest, but usually the least useful. As the name implies, points are randomly removed from the point cloud until a defined target value is reached. The target value is set by the user as either a desired number of points after sampling, or as a percentage of the total number of points before sampling. Random sampling will not change the relative density of points among different areas of the point cloud. If one region had ten times as many points per unit area as another before sampling, that region would still have ten times as many points per unit area after sampling (Figure 5.14).

If left unchecked, large, abrupt variations in point density across the surface of an object can cause trouble in reverse engineering. This problem can be resolved by *Uniform or Grid sampling* (Figure 5.15). Specialized algorithms are designed to decimate point clouds to generate surfaces with an evenly spaced grid of points. There are three common methods to set a target for sampling: absolute number of points after sampling, percentage of points before sampling, and desired point-to-point distance after sampling.

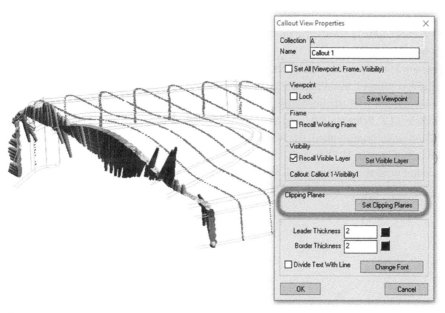

Figure 5.15 Uniform grid sampling is programmed in the software using intersecting planes at defined intervals to capture data on a specific cross section area. Planes can also be clocked around a center point origin when measuring radial components
Source: Image courtesy of Hexagon Manufacturing Intelligence – Spatial Analyzer.

In general, if you double the point-to-point distance, you will reduce the number of points to about one-fourth of the points before sampling. One of the pitfalls of uniform sampling is a higher chance of changing the shape of the point cloud, especially by rounding off edges and corners.

Curvature-based sampling reduces the overall number of points while balancing point reduction and shape definition. The idea behind curvature-based sampling is that one can adequately describe a flat surface with just a few points per unit area (larger point to point distance, or low point density). However curved surfaces require more points per unit area, especially in areas where the shape bends around a small radius (smaller point to point distance, or high point density). Higher point density near tight radii means the sample will preserve the original shape. Lower density flat areas often end up with nearly a grid of points. The software preserves smooth transitions between high-density and low-density areas.

The cleaned and sampled point cloud is now ready for more specific operations, whether it be reverse engineering or quality analysis.

REFERENCES

Anton, H. (1984). *Calculus with Analytic Geometry*. John Wiley & Sons, Incorporated.

Bell, S. (2001). A beginner's guide to uncertainty of measurement. *Measurement Good Practice Guide* 11 (2): Centre for Basic, Thermal and Length Metrology National Physical Laboratory.

Hexagon Manufacturing Intelligence - Spatial Analyzer (2022). Instrument alignment in SA continued: bundling & USMN. https://www.kinematics.com/about/ newsletterarticleinstrumentalignmentcontinued.php. (accessed 8 May 2022).

Kiemele, M.J., Schmidt, S.R., and Berdine, R.J. (1997). *Basic Statistics: Tools for Continuous Improvement*. Air Academy Press.

Weyrich, T., Pauly, M., Keiser, R. et al. (2004). Post-processing of scanned 3D surface data. *Eurographics Symposium on Point-Based Graphics*. https://lgg.epfl.ch/ publications/2004/weyrich_2004_PPS.pdf. (accessed 4 January 2022).

Chapter 6

Reverse Engineering for Industrial and Consumer Applications

6.1 INTRODUCTION

Reverse engineering is the process of examining or measuring an object, and accurately documenting it in three dimensions. The resulting information can then be used to reproduce, upgrade, or provide repairs to that object. It can also be used to establish a baseline model to which future measurements can be compared in order to detect any changes in a part, assembly, or installation.

The documentation most often takes the form of a CAD model, but drawings, or even tables of data could be the end result.

The process can also be used on large objects or buildings; the documentation obtained is useful in a variety of ways, such as maintenance, renovation planning, or replacement. The next chapter describes those applications.

3D Scanning for Advanced Manufacturing, Design, and Construction, First Edition.
Gary C. Confalone, John Smits, and Thomas Kinnare
© 2023 John Wiley & Sons, Inc. Published 2023 by John Wiley & Sons, Inc.

6.2 INDUSTRIAL APPLICATIONS

In industrial and consumer applications, reverse engineering is the process used to create design documentation from an existing object. Because scanning technologies quickly gather data on practically 100% of a part's surface, laser scanners are often the instruments of choice for reverse engineering. Handheld or arm-mounted scanners are most commonly used since most objects for industrial and consumer applications are generally smaller in size. Photogrammetry, structured light scanners, and tracker-based scanners are also, but less commonly, used. Long-range scanners are seldom used.

The needs for reverse engineering are varied. One reason is to produce CAD models or drawings in order to build a part or assembly whose original design information has been lost. This often happens with older designs where the original manufacturer has gone out of business. Sometimes reverse engineering is needed because the design documentation is inadequate to accurately produce a part or a tool or even a mold that made the part (Figure 6.1). Again, this often happens with older designs where parts were made from tooling that did not conform to the original drawings, but assembly adaptations produced parts that worked. The original toolmakers may have discovered interface issues and solved them without notifying engineering. Sometimes it's required due to updating; modifications are planned, and those modifications must fit the as-built shape of the existing parts. An example of this is an aircraft antenna fairing that must fit smoothly and without gaps to an aircraft fuselage. The portion of the fuselage involved must be reverse engineered to provide a CAD model of the fairing's interface.

Figure 6.1 Scanning an antique motorcycle gas tank: the initial step in creating 3D documentation capable of reproducing this item
Source: Image courtesy of Brian Charles Vanderford.

That area of the fuselage must be scanned to provide a CAD model of the fairing's interface per the "as-built" condition rather than the existing CAD models. This will assure an accurate fit in the event that there was any surface deformation during manufacture or in service. Causes of variation from original design are numerous. Often when dealing with aircraft skins or sheet metal and composite formed components, the parts are deformed during installation in order to fit them to the structure on which they are mounted. Another example would be deformation due to normal (or abnormal) use. Frequently aircraft and marine vessels will vary tremendously from the original design after they have been commissioned due to stress, temperature fluctuations, and other environmental conditions.

One can expect that most similar structural assembles will often differ from the original design. However, since the implementation of CAD designed assembles, variations from design have been reduced. Not too long ago, everything was designed on the drafting board along with all of the electrical, hydraulic, and fluid line components. Once engineers were able to design an aircraft fuselage in CAD along with all of the support systems, the likelihood of a first-time perfect fit was increased. Once all of these components were able to be designed in CAD, and the ability to send electronic documentation of these components directly to the CNC machine, tube-bender, or production facility we were able to produce components that were a precision match to the design documentation of the actual assembly.

It is important to note that all design documentation and the fabrication of the parts can be scaled to the correct thermal condition. CAD modeling has allowed for thermal compensation of components during the manufacturing process. When building large aluminum parts, or assemblies of dissimilar materials, thermal compensation is critical. As a metrologist and an engineer, it is of the utmost importance to note that all blueprints and drawings are drafted to represent parts and assemblies at a thermal state equivalent to 20°C or 68°F. Today's modeling, machining, and data acquisition software has the sophistication to take into consideration the material, thermal coefficient of expansion, and the temperature and scale appropriately.

When modeling parts that may have experienced deviation from the design, we call this "as-built" documentation. As-built is the term used in the industry to describe a part or build in its existing condition. In other words, it defines the part or structure as it exists in its current state rather than as it was originally designed. When performing as-built documentation it is again important to note the type of material and the temperature of the part when the data was collected. This can then be entered into the software for scaling or can be scaled by hand using the CTE (coefficient of thermal expansion) of the material. Some of common engineering materials, their corresponding CTE, and the growth expected in a 10–foot or 3-meter-long part that warms up by 40°C, are shown in the table below.

Reverse engineering can be used to create a CAD model that will be used in analysis – stress, fluid flow, or heat transfer. We see this in applications where the as-built shape of the object is more relevant to the analysis results than the designed shape. The Roller Coaster case study at the end of this chapter is a good example of this application.

Table 6.1 CTE Table of Common Manufacturing Materials

Metal	Coefficient of Thermal Expansion μ in/in.°C	Expected Expansion of a 120 inch sheet* (in)	Expected Expansion of a 3 meter sheet* (mm)
3003 Aluminum	23.2	0.11	2.79
5005 Aluminum	23.8	0.11	2.79
6063 Aluminum	23.4	0.11	2.79
Copper	16.8	0.08	2.03
Gilding Metal	18.1	0.08	2.03
Commercial Bronze	18.4	0.08	2.03
Jewelry Bronze	18.6	0.08	2.03
Red Brass	18.7	0.09	2.29
Cartridge Brass	19.9	0.09	2.29
Yellow Brass	20.3	0.09	2.29
Muntz Metal	20.8	0.09	2.29
Architectural Bronze	20.9	0.10	2.54
Phosphor Bronze	18.2	0.08	2.03
Silicon Bronze	18.0	0.08	2.03
Aluminum Bronze	16.8	0.08	2.03
Nickel Silver	16.2	0.07	1.78
Iron	11.7	0.05	1.27
Steel	11.7	0.05	1.27
Cast Iron	10.5	0.05	1.27
304 Stainless Steel	16.5	0.08	2.03
Lead	29.3	0.13	3.30
Monel	14.0	0.06	1.52
Tin	23.0	0.10	2.54
Zinc - Rolled	32.5	0.05	3.81
Zinc-Cu,Tn Alloy	24.9 with grain	0.11	2.79
Zinc-Cu, Tn Alloy	19.4 across grain	0.09	2.29
Titanium	8.4	0.04	1.02
Gold	14.2	0.05	1.27

Most infamously, reverse engineering can be used to steal a competitor's design. This happens less often in practice than most people imagine because reverse engineering is usually expensive and time consuming. It also doesn't carry the full engineering knowledge behind a design. Without taking into account the entire object or designed stress allowances, reverse engineering only one part can't indicate what the tolerances on the part should be.

Successful reverse engineering depends on a clear definition of the project, a suitable object for scanning, appropriate instruments, and adequate post-processing software. Many times, customers say they "want a part scanned." They think that's all that's required, but they are only at the beginning of the process. There are several decisions to be made.

6.2.1 First Decision: Design Intent or the As-built Condition?

Design intent models try to reflect the geometry that the original designer intended. It is often the correct choice when the customer wants to alter or update an existing design or intends to reproduce many parts. Usually at this point, the customer must also specify the CAD software that they want the file to be produced in. CAD software and file extensions are often proprietary and fully featured editable models are usually not easily transferable between CAD systems.

Often the scanned part is imperfect, and the reverse engineer must make a number of decisions about which features are important. Is the slight curve in a surface's point cloud intentional, or should the surface be modeled flat? Or perhaps two planar surfaces are nearly square to each other; should they be modeled as perpendicular, or at a small angle from perpendicular? Should a part that measures 5.002″ long in the point cloud be modeled at 5.000″? (Figure 6.2).

As-built, or as-found, models try to reflect the part as it existed at the time of scanning. If the customer does not plan on altering the part, as-built models can be constructed with surface patches in many point cloud handling software systems. Surface patches are portions of an object's surface that is bounded by a closed curve. The resulting model output is a surface or solid composed of many small surface patches that are often rather randomly oriented. Because this kind of output is not easily edited in CAD it is often called a "dumb solid." As-built models can also be produced as fully featured CAD. Once again, the designer will be making decisions about how to model, but in this case, they are more likely to model that slightly curved surface as curved, and those planar surfaces slightly out of perpendicular. Fully featured as-built models can take a lot of time to complete.

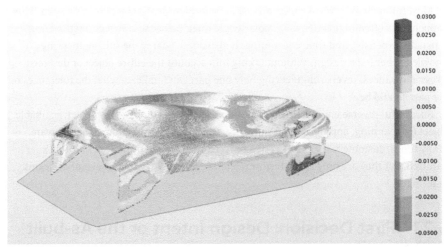

Figure 6.2 A point cloud model illustrating the various fluctuations to the object's existing surface

6.2.2   Second Decision: What Level of Accuracy Is Required?

An important component driving the reverse engineer's decision process is the fidelity of the model to the measured point cloud. Customers will blithely ask for .001″ accuracy and think they are talking about how close the scanned model is to the actual part. From the reverse engineer's perspective accuracy comes in two parts.

The first part of accuracy is to understand that the CAD model won't conform exactly to the point cloud that it was produced from. The software that produces surface patch models is very good. In many cases it will produce surfaces that are within .001″ of the point cloud data, but there will always be areas where the model deviates from the cloud. Corners and small features can be rounded over. Noise in the data shows up as deviations between the model's shape and individual points in the cloud. Furthermore, if the engineer is idealizing the part in any way, by making nearly flat surfaces in the cloud flat in the models, or near-cylinders cylindrical, or nearly parallel features parallel, the deviations will be larger.

The second part of accuracy is that most scanning equipment has a designed precision level of .001–.005″ for small volume parts. There is nothing the reverse engineer can do about this except pick the best equipment possible and explain its limitations to the customer. Part of that is explaining that we can only compare our model back to the measured data and not directly to the original part.

6.2.3   From Point Cloud to a Structured Mesh

One drawback of point cloud files is that the file type cannot describe a surface in a closed solid format. The files must typically be used within a CAD or modeling software as a means to create a solid surface model to be used for reproduction or machining. Sometimes the customer's goal is to skip this step and make parts directly from the model by additive manufacturing (3D printing) or machining. This can be accomplished most easily using laser scanning equipment and software that have the capability to "scan to mesh." In this mode a polygonal surface mesh consisting of thousands or millions of flat triangular surfaces is created during measurement instead of, or in addition to, a point cloud. The polygonal mesh can be cleaned up and sent directly to CAD modeling software or CNC programming software like MasterCAM. The completed dataset will allow the CNC machine to reproduce the part by multiaxis cutting of metal or plastic, or in some instances cut a model to be used for part fabrication (Figure 6.3).

Mesh files are often in STL (STereo-Lithography) format. The file extension for these files is ".*stl.*" Other interpretations of STL are Standard Tessellation Language or Standard Triangle Language. The STL file format has become a de facto standard for many additive manufacturing systems; STL meshes can also be used to create tool paths in many Computer Aided Manufacturing (CAM) software packages.

The STL format goes back to 1987 when it was invented by 3D systems for use with the original rapid prototyping machines. STL files only contain the information necessary to describe a surface body's geometry in terms of the coordinates of the corner vertices of the triangles and a facet normal that conveys the direction of the surface's exterior. No texture or color information is carried in an STL file. Most CAD and scanning software packages can read and write STL files.

STL users should be aware that the file format does not communicate the file units. When using STL files to communicate surface shape, the user must be careful that the export and import units are the same. Tragically, it is easily possible to export an STL in millimeters from the measurement software and import it into a CAD or CNC program

Figure 6.3 A CNC machine cutting an intricate metal part

in inches. The resulting part would be 25.4 times larger in the new software than the original measured part. One must take care when transferring files from one software to another to avoid scaling errors.

STL files are often difficult to modify, requiring the use of specialized programs or point cloud handling software. Since STL files approximate smoothly curved surfaces with flat triangles, parts made using an STL file will have flat facets. To improve fidelity to the original shape the intol/outtol parameters are usually reduced during file export. These parameters set the tolerance for deviations into and outward from the surface. This results in STL files that more closely follow the shape of the part, but at the cost of greatly increased file size.

Other common file formats related to STL are:

Object files: A file format with *.obj* file extension from Wavefront. Defines geometry like an STL but includes color and texture information.

3D Manufacturing Format: *.3MF* files. An open-source standard for 3D file manufacturing.

PLY (Polygon File Format or Stanford Triangle Format): *.ply* files convey information similar to STL, with color and texture capability.

6.3 CASE STUDIES

The following case studies are examples of applications using 3D laser scanning and measurement technology in the industrial sector. They illustrate the diverse and widespread use of metrology within advanced manufacturing and research facilities.

6.4 RACING YACHT BOAT BUILD

6.4.1 Project Scope

In downeast Maine, boat builder Hodgdon Yachts was tasked with constructing a 100-foot racing yacht New^3 (also known as New Cubed.) It had to be constructed and shipped to Australia for its first race within a year's time. Hodgdon's project manager and coordinator explained, "We spent a lot of time planning the construction of each part to ensure everything is built in time for its final installation into the hull or onto the deck." Part of that process was utilizing laser scanning for precise metrology measurements during multiple phases of the hull plug construction. By understanding the exact form and shape of the hull and deck parts could be fabricated to fit precisely, saving time during the construction process (Figure 6.4). This helped to ensure the composite yacht maintained a high level of quality while remaining as lightweight as possible.

Figure 6.4 The ship's hull hung, allowing access to all areas of its hull for scanning

6.4.2 Scanning Process

There were three big obstacles to the shipbuilding project: Time frame, supplies, and skilled labor. The time frame for ship construction is short; suppliers need to deliver boat building materials in a timely manner and boat builders with composite (honeycomb carbon pre-impregnated with epoxy laminate) construction experience are needed. Considering these fabrication hurdles in the initial stages of the planning, Hodgdon Yachts had to coordinate part design and lead times with suppliers to maintain their project deadline. In the case of the New Cubed honeycomb core, there was a lead time of up to 20 weeks for manufacturing and delivery. Advance coordination and project management techniques were key to success.

Bringing a team of scan technicians and equipment on-site to capture the hull plug shape and measure precise distances between end points and mast head, coordination of part design became easier, with less concern of proper fit during installation and fewer materials needed. The technicians worked on-site at the fabrication shop, in tandem with the boat builders. The use of a laser tracker allowed highly accurate hull shape data to be taken despite the confined work area.

During the preassembly stage, laser scanning was used to inspect individual hull plug pieces to ensure all pieces were within tolerance. The surface of the hull plug was scanned, collecting real-time 3D dimensional data, to define whether or not the shape of the hull harmonized with the hull design intent (Figure 6.5). A technician checked for flexing in the bow, verifying key dimensions relative to the drawing. Measurement tolerances for racing yachts allow for little if any building errors, so any deviations from the design intent were corrected or at least communicated to the customer. The laser

Figure 6.5 Surface patches on the hull marked and scanned

tracker was positioned around and on the male mold to check angular distance and form of the hulls unique shape (Figure 6.6).

Figure 6.6 A laser tracker set on the hull to record specific critical measurements

Figure 6.7 Aligning sections of the hull using the laser tracker at the far end

6.4.3   Deliverable

The data provided by laser scanning was given on a real-time fast track basis. This allowed adjustments to be made in the field with multiple follow-up measurements taken to insure proper adherence to design tolerances (Figure 6.7). Laser scanning's precision and accuracy allowed Hodgen to check CNC milled parts to ensure they matched the design specifications before installation. The rapidity of the laser scan process was also a benefit. The scanning of the hull plug reduced what could have been many days' worth of work into only a few hours. A similar process was used for the hull mold setup and final mark out of the yacht's entire structure. This clarification and quality control process of proper measurements ensured that the hull would not crack during the race, providing peace of mind for the racing team.

The Hodgdon Yacht team had a motto "One: Plan well, Two: Choose a good team of people around you with years of building experience, Three: Use good metrologists." They understood that a boat has such a complex shape that metrology was the only real way to check the surface accurately and quickly.

6.5   REVERSE ENGINEERING: CESSNA CARAVAN PERFORMANCE UPGRADE

6.5.1   Project Scope

Engineers at StandardAero, a Dubai Aerospace Enterprise (DAE) company, were asked to increase the performance capabilities of the Cessna Caravan. The project goal, using the current configuration of the aircraft, was to replace the PT6-114A engine with a new increased horsepower engine similar to the PT6. The end result would increase the mission capabilities while allowing the aircraft to carry heavier payloads and fly at higher altitudes. It was necessary to have an accurate 3D CAD model made of the existing conditions. This would provide a clear understanding of the interior envelope available for the new engine, as well as all the various wring, piping, and appurtenances that would require connections.

6.5.2 Documentation Process

Highly skilled metrologists were needed to quickly and efficiently scan the Cessna Caravan to create a dataset of high precision measurement data. The project began with the measurement and reverse engineering of all the aircraft components from the firewall forward (Figure 6.8). Many components were removed, and bench scanned; larger areas of the aircraft were scanned in place while inside the factory. Multiple scanners were used include articulated arms and laser trackers. This data required processing, alignment, and reformatting so that it would seamlessly convert into 3D modeling software.

6.5.3 Deliverable

Once scanning and proceeding were competed, a 3D solid model of the existing engine envelope enclosure was developed (Figure 6.9). From there, the new engine CAD model was fit into place and the design phase began. All components (air ducts, structure, cowlings, etc.) needed to be adjusted for fit and increased air flow (Figure 6.10). Some components required minor redesign. The 3D prototypes of those parts were checked for tolerance and fit using laser scanning equipment.

As a globally recognized aerospace corporation focusing on aircraft leasing and MRO (maintenance, repair, and overhaul,) StandardAero relied on state of the art 3D laser technology to capture the high precision measurement data that would translate

Figure 6.8 The front cowling area to be scanned on the Cessna airplane
Source: Image courtesy of East Coast Metrology.

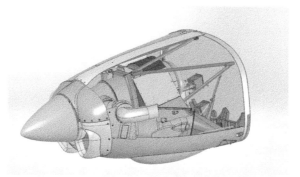

Figure 6.9 CAD model of the cowling revealing the engine inside
Source: Image courtesy of East Coast Metrology.

Figure 6.10 CAD model of the interior area within the cowling, showing placement of both existing and new equipment
Source: Image courtesy of East Coast Metrology.

into a precise solid CAD model for the Cessna Caravan. The CAD-to-part comparison was exactly what was needed, allowing Caravan operators "improved efficiency" and confidence in the redesign of the engine.

6.6 REPLICATION OF A CLASSIC PROPELLER DESIGN

6.6.1 Project

Proper equipment is crucial to commercial fishermen. The owners of Atlantic Trawlers understand this well. Fishing off the Gulf of Maine and George's Bank requires durable boats with strong propellers that fight the tides and get the Atlantic Trawler team to the fishing grounds and back to port safely.

Figure 6.11 The fishing vessel Harmony heading to dock for scanning of the propeller

Atlantic Trawlers has a commercial fishing vessel named Harmony that outperforms other boats the same size. It's known as one of the top fishing vessels in the region. The boat's fair lines, powerful engine, and propeller allowed it to efficiently harvest fish in vicious tides (Figure 6.11). The propeller was one of the key elements separating that boat from the rest of their fleet. Other boats of similar shape and engine horsepower lacked the performance of Harmony. Unfortunately, one drawback of older ships is they lack drawings identifying design dimensions. The propeller was over 25 years old; there were no original drawings or molds. Thus, Atlantic Trawlers was faced with the challenge of recreating the four bladed propeller dimensions if they wanted to replicate it for use on other ships in their fleet.

6.6.2   Scanning Process

The ultimate goal was to develop the CAD model so Atlantic Trawlers could manufacture both a back-up prop for Harmony in case it was damaged and to provide similar props for the rest of the fleet. Compounding matters was the limited access to the propeller, available only during a brief haul-out of Harmony for routine maintenance. Use of 3D laser scanning for data collection would be a quick, accurate, and effective method to document the existing propeller, and would not delay the return of the vessel to the commercial fishing grounds.

Figure 6.12 The laser tracker set in place scanning the propeller

A laser tracker was used to measure the propeller; all measurements were completed within just a few hours. The main challenge in gathering the measurements was the propeller, which was mounted on the ship's drive shaft under the hull. The laser tracker, with its ability to scan up to 1,000 points per second, proved to be ideal for the task. It is sufficiently portable to fit in the confined area under the hull and allows the technician to move freely around the propeller as the machine tracks the points (Figure 6.12).

6.6.3   Deliverable

The propeller's measurements were loaded onto a laptop and taken back to the metrologist's office. There the data was imported into a CAD package that allowed constructing features and surfaces from the point data. To construct the new propeller model, the point data from one of the four blades was converted to a NURBS surface. NURBS is a mathematical model composed of splines that describes complex surface geometry. The other three blades were then created by rotating that CAD rendered blade 90 degrees around the propeller center until there were four identically shaped blades in the model. As a quality check on the process, the actual measurements for the other three blades are compared to the "engineered" surface of the main blade to look for disparities and any deformation that needs to be considered.

Figure 6.13 shows the point cloud data for the propeller along with the constructed reference frame located at the propeller center and clocked through the reverse engineered propeller blade. The other three blades show the density of the point data collected to ensure adequate coverage for assessing symmetry to the original blade.

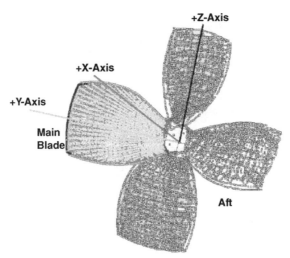

Figure 6.13 Point cloud dataset formed by the scanning process, with the propeller axis overlaid for proper orientation

Once the surface was established and all four propeller sections in place the model was converted into STL and IGES solid surface format CAD files. These were delivered to Atlantic Trawlers for their use in generating CAD-based drawings and surface definitions. Those digital 3D models could be used to create a mold to pour a replica of the Harmony propeller or to generate a five-axis CNC machine program to mill the propeller from a block of metal alloy.

The ability of the 3D laser scanning process to collect propeller data within a quick time span, generate a 3D CAD model, and provide a quality check on the existing propeller offered a distinct advantage to Atlantic Trawlers over traditional time-consuming reproduction methods.

6.7 ROLLER COASTER INSPECTION

6.7.1 Project Scope

When it comes to amusement park attractions, roller coasters seem to top the list of thrill rides to experience during your visit. But, when you finally step onto that precision engineered attraction, do you think about how it was designed, constructed, or inspected? Like any other large structure, whether a bridge, a plane, or skyscraper, there is a lot of engineering work behind the scenes that make it safe for people to enjoy on a daily basis. And safety is an amusement park's number one priority!

For the past 35 years, Dr. Masoud Sanayei, a Professor of Civil and Environmental Engineering, has conducted research on structural health monitoring of bridges and other structures. By using instruments such as strain gages, accelerometers, tiltmeters, thermocouples, optical sensors, and various data acquisition systems, he is able to collect, measure, and process information that enables him to comprehensively evaluate the condition of these structures. His computer models of existing structures can be subjected to a series of loading patterns within finite element simulations for prediction of structural system responses. The results of these simulation models display how a structure reacts to repetitive cyclical loading from a fatigue perspective. This data can then be used to isolate specific regions of a structure that would benefit from fatigue loading studies for potential preventive maintenance.

To determine the viability of structures, one must create a physics-based mathematical model to interpret and evaluate the predicted response of the structure compared to its actual measurements. In order to create his diagnostic computer model, Professor Sanayei needed to have the exact geometry and material properties of the roller coaster structure. This would require accurate as-built documents of the roller coaster structure. While most modern roller coasters have digital design files, older structures might not have such drawings. The roller coaster chosen for his study did not have the documentation needed to conduct the research. Another important factor to consider was settlement. After the ride has been constructed, it has a tendency to settle, creating additional areas of uncertainty. By producing an as-built model of the current conditions the simulations would accurately represent the conditions in the field.

A metrology firm was selected to provide the 3D laser scanning and fine measurements needed to produce as-built documents. Scanning was also a preferred method due to its speed of data acquisition, allowing less interference with the amusements parks patrons. The metrologists were tasked with providing an as-built model of the roller coaster which could be used to run mathematical models and simulations. The model was to be composed of all the components: the foundation and main support columns, all connections to the box girder beam, support brackets, and the tubular rails.

6.7.2 Scanning Process

The scope of the project focused on measuring and modeling four box girders and three supporting columns with their attachment gussets, the plates that anchor the roller coaster. When choosing which scanners to use for the job, several variables had to be taken into consideration such as size of the structure, area of interest for scanning, distance to the target structure, weather conditions, movements, and accessibility. Outdoor structures that are extremely high can also be subject to temperature and wind variables. The FARO Focus 3D Laser Scanner was selected as a means to efficiently capture a widespread area of the roller coaster's first downfall section (Figure 6.14). Once the required regional data was captured, additional close-up hand scanning was taken using the FARO Edge Portable CMM Arm (Figure 6.15). This scanner provided a precise cross-scan of the welded sections. When combined with the terrestrial laser scans, a comprehensive digital geometry model of the high stress areas within the research scope was achieved.

Figure 6.14 Roller coaster being scanned by the FARO lidar scanner

Figure 6.15 Sections of steel supports scanned at high resolution by the FARO arm scanner

Heavy duty magnetic bases were used to assist in securing the equipment to the edge of the coaster. A scissor lift gave the equipment visual access to the upper reaches of the roller coaster. The goal was to acquire as much data as possible in a relatively short period of time with minimal noise (excess data that is irrelevant). By careful selection of scanning positions and heights from a range of 30 meters, a 360-degree coverage of the target sight was acquired.

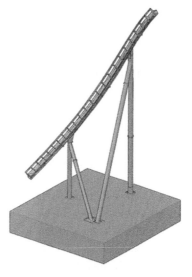

Figure 6.16 Image of the roller coaster track and CAD model of track derived from on-site scanning data

6.7.3 Deliverable

Once the scanning phase was completed, the post-processing phase commenced. This consisted of stitching all the individual scans to one another to create an aligned 3D digital file of the roller coaster. Superfluous data such as ground cover, surrounding trees, other structures, and noise was removed, and the resulting file converted to a polygonal mesh. The final step was to use the mesh model as a basis for creating solid surface geometry of the structure of the coaster. This produced an "as-built" digital model, a full parametric model representing the collected metrology data (Figure 6.16).

Additional sensors and monitors had been installed at critical positions on the roller coaster. The data from those instruments was combined with the digital model produced by laser scanning. Pairing the measurements taken from the sensors and monitors with the data collected from 3D scanning provided valuable information on the real-time effects of fatigue and stress on the structure. The study would allow for multiple sections of the roller coaster to be accurately monitored for stress moving forward, allowing better long-term maintenance planning.

6.8 ARGOMAIDEN SCULPTURE

The development of laser scanning as a means to copy intricate geometric shapes has led to its widespread use by many sculptors. Reverse engineering is often used as an initial step in creating artwork, either for reproducing a limited run of pieces based on the original, or as part of the process involved in creating a single piece of art. Laser scanning provided a crucial role in the development of *Argomaiden*, a 3D wood sculpture that is the result of a collaboration among three Newburyport, Mass., artists; Rob Napier, Gordon Przybyla, and Eric Smith. At first glance, the artwork appears as a stylistic fusion between a half-hull model of a nineteenth-century sailing vessel and a shapely mermaid figurehead with long tresses. Half-hull ship models are wooden models used by ship builders of the past to design in three dimensions the hull forms they would build. A ship's figurehead embodied the vessel's spirit and was originally believed to placate the gods of the sea and ensure a safe voyage. Although figureheads are not typically part of half-hull models, *Argomaiden* exaggerates and combines these forms into a surreal notion of a classic half-hull ship model.

The original artwork had been formed at actual scale, crafted from a combination of wood and artist's clay. The intent was to use the artist's model as the pattern for the final piece of art. Ultimately, the four-foot-long by one-foot-wide final piece would be carved in laminated mahogany, given a fine wood finish, and wall mounted. To achieve the final form, scanning of the model was required to provide digital data that could be used to establish a cutting pattern for a CNC machine. The resulting cut wood sculpture would then be hand finished by the artists as needed.

In preparation for scanning, the scan technicians smoothed the model's rough surfaces to hide imperfections such as material interfaces and surface defects while preserving important artistic details around the figurehead's hair and profile. The model was reverse engineered using a laser scan tip mounted to an articulated Faro Arm. The laser scanner gathers accurate surface measurements in the form of point cloud data without physically contacting the surface. This was an important part of the process, as any contact with the smooth soft clay finish of the model during

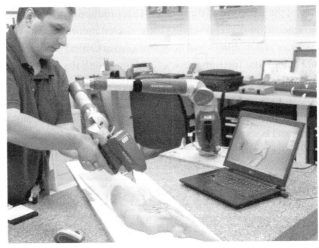

Figure 6.17 Scanning technician using an arm scanner to record the surface of the artist's sculpture

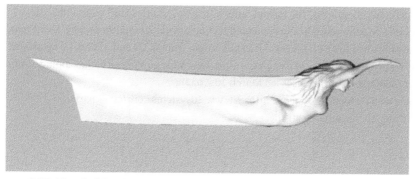

Figure 6.18 The finished digital 3D solid surface model of the half-hull sculpture

scanning could negatively impact the object's intended design. During the scanning process the technician was able to view the collected scan data on a laptop connected to the scanner (Figure 6.17). This 3D view of the scan data while in progress insured that he was able to capture all areas of the sculpture's surface. Once all scanning was completed the raw data was brought into Geomagic software to generate a surface mesh CAD file from the laser scan point cloud data. In the modeling software the mesh model was further processed to its final digital form, a watertight solid surface model (Figure 6.18). This smoothed surface model was then used to drive a three-axis CNC milling machine to form the artwork from a laminated block of mahogany. Following the rough machine cutting, hand sanding and finishing completed the piece, making it ready for display.

REFERENCES

ECM – Global Measurement Solutions (2012). Reverse engineering the Cessna Caravan's performance. https://eastcoastmetrology.com/resources/applications/reverse-engineering-the-cessna-caravans-performance. (accessed March 3, 2022).

ECM – Global Measurement Solutions (2016). Better boat building starts with Pre-planning and precision metrology. Quality Digest Article.

ECM – Global Measurement Solutions (2020a). Remaking a classic propeller in quick-turn time. https://eastcoastmetrology.com/resources/applications/remaking-a-classic-propeller-in-quick-turn-time. (accessed March 3, 2022).

ECM – Global Measurement Solutions (2020b). Taking roller coaster accident prevention and maintenance to the next level using 3D scanning technology.

Erik Oberg, F.D., Jones, H.H., Ryffel, H., and McCauley, C. (2020). *Machinery's Handbook: Toolbox*. Industrial Press. (accessed March 30, 2022).

STL (STereoLithography) file format family. In 1986, Charles Hull co-founded 3D Systems to commercialize his patented technology for stereolithography. He developed and published the STL file format to allow data from CAD software to be translated for 3D printers. *Library of Congress.* https://www.loc.gov/preservation/digital/formats/fdd/fdd000504.shtml. (accessed March 30, 2022).

What is an STL file? *3D Systems.* https://www.3dsystems.com/quickparts/learning-center/what-is-stl-file.

Chapter 7

Reverse Engineering for Architecture, Engineering, and Construction

7.1 INTRODUCTION

While reverse engineering is commonly associated with industrial applications, the principles of reverse engineering can also be applied on a much larger scale to the built environment. Objects such as buildings, bridges, and other large-scale structures benefit from the use of 3D laser scanning documentation. By using laser scanning to accurately document existing conditions of a structure, or part of a structure, a digital document is formed to become the basis of analysis for study, planning, or replacement. Often this is done in a noninvasive way, thus saving time and expense for the owner.

3D Scanning for Advanced Manufacturing, Design, and Construction, First Edition.
Gary C. Confalone, John Smits, and Thomas Kinnare
© 2023 John Wiley & Sons, Inc. Published 2023 by John Wiley & Sons, Inc.

7.2 MODELING FOR ARCHITECTS AND ENGINEERS

Industry, in conjunction with the guidance of architects and engineers, has begun to adopt the use of 3D scanning to create digital twins of entire facilities. This has been made possible by the advances in scanning, software, and computer technology alongside the widespread acceptance and use of 3D CAD (computer-aided drafting) and scanning within the architecture and engineering professions. As design moved from 2D to 3D documentation, scanning provided a means to capture existing conditions of buildings and equipment in a 3D format. Scanning's unique ability to accurately record large areas of both flat and complex geometries has made it a valuable tool within the AEC industry. When properly scanned in the field, the resulting data will capture every nuance and detail of a building's surfaces, furnishings, and equipment. As modern design leans toward creating buildings with multiple undulating surfaces and nonplanar walls, and renovations are needed for centuries-old masonry buildings that have begun to deteriorate and shift, 3D laser scanning becomes a desirable and effective method to gain accurate documentation for both design and construction.

Advantages of 3D laser scanning:

Why is 3D laser scanning used for data acquisition within the AEC professions? Among the advantages it provides are:

- Scan accuracy of +/−2 mm at 100 feet distance
- Creates a 3D digital record of a 360 degree area within minutes
- Reduces time in the field and return trips to the project site
- Enables measurement of building elements and details without need for scaffold or ladders
- Measures hazardous or unapproachable elements from as far away as 500 feet
- The data serves as a basis for 3D BIM (Building Information Modeling) and 2D CAD models
- Multiple scans aligned create easily shared digital models for reference and study

Other qualities of scanning are more specific in their application but just as important. It can serve as an active component in enabling sustainable design goals. Scanning a building provides precise volumetric and square footage data for energy modeling studies. It accurately records in detail the complex shapes and forms of older ornamentation, allowing a new replacement piece to be created before the old one is removed. Timber structures, along with their joinery, can be easily documented in place, providing noninvasive replication for preservation and adaptive reuse. Just about any aspect of the design and construction process can be aided by the features of quality 3D data.

7.2.1 Applications with the AEC Industries

3D laser scanning has become an accepted method for documenting existing building and site conditions. It provides a fast, accurate, and comprehensive means to obtain a 3D digital image of a building or structure. This digital data can be imported into 2D and 3D cad programs for creating architectural and engineering drawings. Once imported this true-scale digital image can serve as a basis for creation of 2D floor plans, sections, and elevations, 3D models, topographic mapping of sites and building surfaces, volumetric studies of land and excavations, or documentation of building sites preconstruction and during construction.

Scanning of a building is usually done within 20–150 feet from the façade or area of the building to be documented. The scanner is mounted atop a stable, level tripod. Resolution, or scan density, should be set to provide a quantity of data points yielding a level of detail sufficient to allow creation of the desired CAD deliverable (Figure 7.1). Scanning is line of sight only, so the distance between setups is not only dictated by the scanner's range, but usually by the need to catch certain elements or spaces by multiple angles of sight. Simple orthogonal interiors or facades can be scanned quickly. For more detailed drawings, such as 3D models with multiple pipes, ductwork, or ornamentation, resolution must be set higher and additional scan station setups are required to gain full visual coverage of the areas. This will increase the field time needed for scanning. Many scanners have a built-in color camera, which provides an RGB value to the otherwise typical black and white grayscale scanning data. The addition of color makes it easy to differentiate various piping and other appurtenances within the building, adding value to the scan data. Some of the uses for 3D scanning within the AEC industry are:

- Building documentation
- Surface and forensic structural analysis

Figure 7.1 Typical view of point cloud data of a building formed by multiple scans at high resolution

- Clash detection
- BIM modeling
- Digital twin creation

7.2.2 Building Documentation

The term building documentation refers to the act of capturing and recording data and measurements that define the shape, size, configuration, and materials of a building and its various components. This data can be used in two ways. The first is using the documentation process to prepare a set of 2D or 3D plans which provide a comprehensive descriptive record of the building. The second is to use the documentation at a component level to prepare actual parts or elements of the building.

Whether a building is scanned by a static ground-based scanner or a mobile scan system, it is important to gather scan data that will serve the needs of your intended documentation and deliverables. Often buildings have to be documented at a high level of accuracy to provide a reliable set of plans for construction. Sometimes dense high-quality scan data is required to record small or complex elements. Other times data needn't be as precise, particularly if its intended purpose is to provide basic floor plans or 3D interior visuals used for schematic planning or reporting purposes.

Figure 7.2 Note level of detail within the point cloud data, allowing for CAD delineation of façade elements

After the point cloud data is captured in the field it is processed and aligned to form a 3D digital file of the building. This data is then brought into a CAD program to aid in the development of 2D or 3D drawing. The aligned data serves as an underlayment for drawing cad lines. Zooming in and out of the scan data within the CAD program enables one to discern various elements, and place lines to accurately denote various building parts such as windows, trim, and ornamentation (Figure 7.2). Specialty software add-ons can automatically begin the forming of walls, piping, and other objects within 3D software. The level of detail obtained will depend upon how well the building was scanned. Dependent upon one's needs and adjustment of the scanner settings, a building scan will produce point cloud data able to discern fine detail such as the actual brick coursing of the facade.

The caveat in these methodologies is that they are dependent upon the skill and expertise of the person drafting the image from the data. Once a line is drawn, it represents one's interpretation of the data. In order to achieve a reasonable level of accuracy, it is important to have a well-processed set of point cloud data and photos as well as personnel trained in processing and understanding the data. Once the drawing is completed it is referred to as an "as-built" or "existing conditions" drawing, as it represents the building's condition and configuration at a specific point in time, i.e., when it was measured and scanned. This drawing file, along with the point cloud dataset, can be used for design planning or archived and held for future reference.

Sometimes the scanning process is limited to specific elements or purposes. Ornamental iron, masonry, and stone contractors often use scanning data to replicate existing items that require replacement. By scanning the item while in place, it insures an accurate fit for the new piece in its surroundings. In the case of exterior ornament, scanning allows the original piece to remain intact while its replacement is being made. This is especially helpful when a project extends through many months, as it eliminates the traditional molding in place process, or removal of an often fragile item that could break apart as well as open the wall fabric to the ravages of weather. The use of digital information also simplifies reconstruction of an element. Scanning data of decorative elements can be converted and digitally modified in the computer, producing a 3D digital model of the intended replacement that all interested parties can view and approve prior to any fabrication.

7.2.3 Surface and Forensic Structural Analysis

One key advantage to using scan data is exploiting its vast amount of data points upon a surface for forensic study of surface characteristics. While construction drawings may be straight and true, real-life construction is typically less precise. This is especially true when dealing with an older building. Walls and floors often have more in common with a sheet of paper rather than a plate of glass, with numerous minor shifts in level, tilt, and surface wrinkles. Scan data can be used to document these anomalies in a topographic style leveling map, providing a clear indication of problem areas all tied to a set "zero-zero" base point.

Walls can be viewed in either plan view with contour mapping or in small cross sections, indicating the outward or inward tilt of a specific area of the building. The scan data can be brought into a CAD program and sliced to create a section of specific thickness. The resulting view will not only show profiles of any trim or molding, but also illustrate deviations from a straight vertical line. When mapping walls, a specific layer cut depth is prescribed, such as ¼″ thick. This depth is then cut into or outward from the wall off a prescribed starting surface elevation point. Each layer can be color coded based upon its assigned depth. When placed together, a clear picture of the wall's surface deviations is presented. This mapping can detect depressions, bulges, or other surface anomalies in the façade. It can indicate if a wall is leaning or tilting off vertical. Identification of these irregularities at an early stage of the project helps to focus design efforts.

Mapping of this type is especially helpful when dealing with a rusticated wall. Traditional discrete spaced survey points will not give a clear picture, as the immediate surrounding wall surface can deviate up to an inch or more due to the rusticated surface cut of the stone. However, full scan data coverage of the wall will average out these deviations, and give an overall picture which clarifies the wall's true condition. Figure 7.3 indicates a rusticated stone wall, with topographic slices at 3/8″ intervals, showing a noticeable tilt outward as the wall rises above grade. This enhanced information allows the professional to focus their study when in the field, or advise the owner as to the possible extent of problem areas and remediation costs.

A similar process can be applied to floors. This aspect of scanning is widely used in the construction industry to calculate the levelness of existing and new floors. Figure 7.4 illustrates a floor plan of a modern office building under construction, with the resultant floor mapping overlaid on the proposed CAD floor plan. Scanning the floors and preparing a leveling map, tied to a base point of an elevator or stairway sill set at elevation zero, yields a clear understanding of the areas that require fill and those that need to be ground down. Having the ability to map those areas within a CAD program also enables accurate

Figure 7.3 Left to right: Point cloud data of wall, section at one specific depth within overall wall surface, "topo" style mapping of entire wall showing various surface levels

Figure 7.4 Left to right: Scanner within open room, note targets in place for accurate alignment of multiple overlapping scans. Final image of floor indicating topography captured by the scanner's high resolution

dimensioning of the affected regions, making it easy to locate them on-site. Quantities can be easily calculated from the data, allowing properly quantified bids. Floor mapping can also provide insight into possible structural issues that may exist within a floor, graphically illustrating depressed areas that have resulted from bending stress or overloading, including the weight from added floor leveling applied by previous interior renovations.

7.2.4   Clash Detection

Clash detection refers to the comparison of existing in-place elements of a building to the intended design elements to verify locations and identify where they interfere with one another. This data is extremely useful during both the design and construction phases of the project. A 3D model which accurately portrays the location of walls, columns, and other interior elements can be formed from point cloud scan data. This plan can be imported into CAD and aligned with a proposed 3D model of the space to see where any deviations occur. When measuring large spaces by traditional hand techniques, overall distances can sometimes be skewed due to compounding of small inaccuracies in a series of linear dimensions. When properly scanned with the use of targets, the overall accuracy of a series of scans will be much higher than that done by hand. Comparing scan images against hand-measured plans will often reveal inaccurate placement of walls and objects.

Existing conditions such as ceiling beams can be located from the scan data; their pitch and actual height above a prescribed elevation point can also be established. Existing ducts and pipes are all located within the space, along with any open penetrations in structure or framing for ductwork installation. All of this information, if obtained during the design phase, will help enable a designer to provide a set of construction documents that will be properly coordinated and allow a smoother flow of construction and equipment installation.

Contractors often have an interior space scanned following completion of demolition work. The resulting 3D data is then clashed against the design BIM models for architectural, mechanical, electrical, and plumbing trades (Figure 7.5). The resulting enhanced BIM model enables them to identify at a preconstruction stage any clash issue, such as new pipes or ducts penetrating a portion of an existing wall or beam. This allows correction of these potential clash issues before fabrication or installation of new construction, saving time and money during the project.

On fast-tracked projects, where construction CAD documents are prepared as construction work progresses, scanning is used to locate completed work, such as foundation and shear walls, steel locations, and floor levels. This information is relayed to the architect or engineer who uses it to complete the next stage of design, confident that the newly prescribed work will fit into place. The scanning data is often shared with various subcontractors as well. Finish installers will use the leveling maps of a wall surface to determine how much play they need to build into their panel adjustment clips, tile installers will use it to determine grout leveling quantities for floors. Cabinetry and wood work installers will use it to ascertain exact wall to wall dimensions and determine if a room's walls are square and plumb.

The use of scanning data allows the construction process to proceed faster, reducing installation clash and fit issues and verifying their work aligns with the construction drawings. As preconstruction scanning data is used by more architects and engineers, it will hasten the design process, providing faster timetables for both design and construction.

Figure 7.5 Clash Detection overlay of CAD model with existing scan data

7.2.5 BIM Modeling

One could argue the development of BIM provided a catalyst for the adoption of 3D scanning technologies in multiple industries. As various disciplines move toward 3D modeling and drawing, having a means to capture 3D digital data at the outset of a project is a tremendous time saver in preparing base drawings for a project. Two of the major CAD platforms, Autodesk and Bentley, have installed point cloud usability directly within their CAD programs. Other third-party software developers have made advances in specialized add-on software to produce programs that readily identify walls, surfaces, and piping within point clouds in a CAD environment. Some even automate the process, rapidly decreasing the time it once took to draw a complex piping network. The increase in software abilities, coupled with 3D laser scanning's proven accuracy for as built conditions, has led building documentation providers to have facilities scanned and forego traditional hand measurement methods. The resultant point cloud, a confirmed dataset of current existing conditions, becomes the basis for a 3D model.

Scanning is most effective when used to document those building types or elements that are difficult if not impossible to scan using traditional hand measurement methods. One example is documentation of a room with numerous pipes and ducts. Renovations require manipulation of current building systems or relocation of piping to accommodate new systems and structures. It is also important to identify what is contained within the pipes and their material makeup. Scanning will allow these existing conditions to be documented, and then brought into a CAD format to produce a 3D model. Using the camera imaging option present in many of today's laser scanners, the CAD operators are given a clear visual picture of the piping, with the color imaging becoming an easy source of reference to differentiate between piping types.

Scanning for piping and building systems documentation is typically a more time-intensive and costly endeavor than when scan data is used primarily for walls and surfaces. It is important to scan at higher levels of resolution when the data will be used for pipe documentation. This enables adequate coverage of data points on the conduit and piping, to be certain all elements are captured. Piping displays best when scanned from more than one side to give a clear picture of the pipe's shape and any deformations. This typically involves more scan station locations within the area to be documented, increasing one's field hours. However, the additional time spent results in a model that can become a base for the work of all disciplines involved; architectural, structural, mechanical, electrical, and plumbing. This increases the cost benefit return, one of BIM's prime benefits.

7.2.6 Digital Twins

Digital twin refers to a virtual representation of an object or facility. The term's definition varies depending upon the sector or business to which it is applied. Typically it is used to describe a 3D model of a facility or object, but the term is really meant to convey much more than a model. It begins with the building's physical form and materials documentation, like that found in a BIM model. What makes it a digital twin is the next two essential elements: the input of ongoing sensor data regarding the building's use and the ability to update the model to reflect the data input and changes to the facility over time.

This use of data, the so-called fourth dimension in 4D models, allows the digital twin to become a powerful and useful format for study of the facility's industrial processes, energy use, and personnel work flows. The digital twin of an industrial facility will allow energy management, inventory controls, ongoing maintenance, and scheduling of shift times to optimize production. Simulations can be run within the digital twin to assess how planned renovations or additions will impact use and efficiency of the facility. Offsite consultants or managers can easily access the digital twin to provide feedback and review of manufacturing workflows or process failures in real time.

Scanning's usefulness in digital twin creation is its ability to provide an accurate as-built dataset of a facility's current condition. Portions of scan data, or scans of equipment, can be inserted directly into the digital model, instead of creating a 3D CAD-based approximation of an object. This not only provides a more accurate depiction of the internal environment, but enables the model to be simply updated over time as equipment and production lines change within the facility. Ongoing scanning of a facility, or use of LIDAR sensors within the building, provides a continuous stream of updated information on the building. Forensic study of updated scan data over the facility's lifespan can yield valuable information foretelling potential dangers from wear or stress on the building's structure, piping, or machinery. Scanning technology will continue to grow in value as the use of digital twins expands within the manufacturing industry.

7.3 CASE STUDIES

The following case studies are examples of 3D laser scanning use in the architectural, engineering, and construction sector. They illustrate applications of laser technology using various pieces of equipment, from highly precise tripod-mounted long-range measurement devices to handheld units. The methodologies used were for a wide range of projects including survey controls, total building documentation, and ornament replication.

7.4 RESEARCH FACILITY COORDINATE CONTROL SYSTEM

7.4.1 Project Scope

A national laboratory dedicated to research of fusion as a safe and renewable energy alternative built a tokamak. A tokamak is a fusion machine that uses magnets to harness superhot plasma and fuse atoms together, which in turn creates energy. The laboratory wanted the ability to accurately align and adjust equipment relative to the tokamak's vacuum vessel. To assist in managing its experiment, the laboratory wanted to characterize magnet geometry and establish a network of permanent control points.

A global coordinate system was to be setup within the facility, used to easily describe 3D spatial relationships between all geometry and control points. The coordinate system was established based upon datum feature information provided by the laboratory. As a result, a coordinate system was established relative to the tokamak vessel.

7.4.2 Coordinate System Requirements

In order to be relative to the tokamak's vacuum vessel, reference features with known locations needed to be defined.The reference features used were 0.5″ and 1.5″ SMR control points. When at least three control points are measured, the measuring instrument aligns itself to the tokamak's coordinate system. The laboratory wanted at least 40 permanent control points on the outside of the tokamak vessel as well as 16 on the inside of the vessel. All permanent control points had to be relative to the tokamak's coordinate system with an uncertainty of under 0.007″. The tokamak is spherical, spans three stories, and has openings to the inside from the top and bottom. Other features were defined relative to the tokamak's coordinate system. These included the magnets, or poloidal field coils, that wrap around the tokamak both vertically and horizontally. The centerlines of the vertical coils and horizontal coils had to be defined within an accuracy of +/−0.015″ or better.

7.4.3 Specialized Laser Scanning Equipment

This unique project called for multiple metrology instruments working together to create one solution. Due to the size of the tokamak and the level of accuracy required, laser trackers were the preferred instrument (Figure 7.6). For this particular application, four

Figure 7.6 Laser tracker set within the tokamak to capture distinct reference control points

different laser trackers where used, each providing unique functionalities. The Leica AT401 and AT402 provided the ability to easily move from one location to the next, due to their lightweight and built-in battery back-up capabilities. The AT401 and AT402 also have the ability to shoot directly up, useful when operating at the bottom opening to the tokamak. The Faro Vantage laser tracker can measure objects in confined spaces, as well as mount horizontally to see into the tokamak from the top opening. The AT960 has the same capabilities as the Faro Vantage, in addition to some special accessories the other trackers lack. One of these special accessories is the Leica Absolute Scanner (LAS). The handheld LAS scanner can acquire large amounts of point cloud data through noncontact measuring.

7.4.4 The Process

To achieve +/−0.007″ location accuracy of the permanent control points on the tokamak's vacuum vessel, a Unified Spatial Metrology Network (USMN) needed to be created. To overcome line of sight issues caused by the size and geometry of the tokamak, roughly 100 tracker positions were set up over three platform levels (Figure 7.7). The USMN feature within the Spatial Analyzer software was used to perform a weighted bundle adjustment of all the trackers, and their measured control points. The USMN feature calculated the uncertainty of each control point and created a nominal point that has greater certainty of its location to its true value. Relation between points

Figure 7.7 Laser tracker set outside the tokamak

that have no line of sight between one another (such as points on opposite exterior sides of the vessel or on different flooring levels) needed to be created by additional tracker positions. In order to make the USMN reach the desired accuracy, multiple line of sight paths to connect points inside and outside of the vessel, on different platform levels, and on opposite sides of the vessel were established (Figure 7.8). The use of multiple station setups and pathways adds more uncertainty to the actual relation of the points. The uncertainty can be minimized by using a different path such as moving clockwise around the vessel if the original path was counterclockwise.

While measurements were being taken for the USMN, other desired features were captured, including the vertical coils and horizontal coils. This task required the use of the AT960 in conjunction with the LAS scanner. This instrument allowed the scanning technicians to quickly measure and obtain enough data to accurately model the features. Once a model was created from the scan data, the centerline of the coils could be easily located and defined. A digital database was provided indicating all xyz points for the control system that had been set in place and verified through laser scanning. This system would now enable future equipment and accessories to be set accurately in place to insure consistent results from the tokamak research.

Figure 7.8 Tracker used in close proximity to record surface-mounted CMR data

7.5 ARCHITECTURAL ORNAMENT REPLICATION

7.5.1 Project Scope

Duke University's main quad is composed of a series of masonry buildings dating back to the late 1920s. The buildings were all designed in the Medieval Gothic revival style. The facades are constructed from ashlar masonry stone, known as Duke Stone, from a local quarry, and are trimmed with moldings, gargoyles, and ornamentation sculpted from limestone. These limestone details lend a character and beauty unique to the campus. Unfortunately, as the buildings age weathering causes limestone, a soft rock, to degrade over time. In an effort to preserve and document some of these unique details, laser scanning was used as a means to create digital 3D replicas of the ornaments. This database of decorative elements would provide an accurate digital model that could be used for study or replication of the objects.

7.5.2 Scanning Process

To properly document the various pieces of ornamentation a hand scanner was required that would have the ability to record the slight surface markings and complex geometry of the objects. A terrestrial ground-based scanner was used for initial scanning of the building to provide context for the various individual elements scanned (Figure 7.9). A Lecia 960 Tracker with a Lecia Absolute Scanner was selected for the process. This is a two-part system, the first piece being a handheld scanning device that emits a visible

Figure 7.9 Tripod-mounted scanner and resulting scan data of building entry

laser beam. As the operator moved the laser light over and around the object, an image of the scanned object appeared on a tablet. This ability to view the object as it was being scanned insured complete coverage of surfaces. The second part of the system, a laser tracker, provided a continuous connection to the handheld unit, recording its precise location in space. Together these devices produced a finely detailed point cloud file of the architectural ornamentation.

Multiple pieces and parts of architectural ornament on the buildings were scanned. Some were within arm's reach at ground level; others required a boom lift to gain the close access required for scanning (Figure 7.10). Once scanning was completed, all datasets were uploaded to a computer for further processing. In the computer the data was opened in 3D modeling software. It was cleaned of data noise, and individual sections of scan data were put together as needed to form the 3D representation of the object scanned. The object was then made into a "watertight" model. This entailed finishing the scan model within the software to provide a complete enclosure of the outside surface of the object scanned, free of any holes or voids. The final model became an accurate representation of the scanned object, and could be used as a basis for making a 3D print or CNC file (Figure 7.11).

Figure 7.10 Use of a handheld scanner to record intricate detail of ornamental objects

Figure 7.11 Processed and aligned scan data of shield as seen within the modeling software

7.5.3 Deliverable

To test the quality of the 3D files, it was decided to produce a few 3D prints. The digital files of a few gargoyles were processed and sent into a 3D printer. The printer produced

Figure 7.12 Completed dataset of scanned object with resulting 3D printed reproduction of the object

a full-scale polymer resin facsimile of the scan model. The model was made using the additive process, with the printer slowly moving along a prescribed pathway laying layer upon layer of polymer resin to form the 3D object. The 3D prints produced a realistic copy of the object, including its tool marks, and could serve as an object for closer study or display (Figure 7.12).

7.6 FONTHILL CASTLE, BRONX, NY

7.6.1 Project Scope

Documenting a rusticated masonry building is often challenging. Add to that a multifaceted, varied building configuration, both interior and exterior, and 3D laser scanning quickly becomes one of the few methods available for capturing accurate dimensions of existing conditions.

Built in 1852, Fonthill is a Gothic Revival stone masonry building consisting of five octagonal towers of varying heights radiating from a three-story central tower (Figure 7.13). In 1980 it was listed on the National Register of Historic Places. Owned by the College of Mount Saint Vincent, the Castle has served multiple uses over its lifetime. When prospective alternative uses for the building were being considered, it was decided to document the building to provide an accurate basis for future renovations and additions.

Figure 7.13 LIDAR scanner outside the building; note spherical targets nearby

7.6.2 Objectives

In addition to capturing the overall size and configuration of the building, information was needed to assess the structural condition of the masonry. The masonry and mortar joints needed repair in a number of places. Multiple cracks seen in both interior and exterior indicated the possibility of structural deterioration. Interior spaces were varied in size and shape, from large vaulted spaces to narrow twisting stairways. Accurately aligning the multiple levels presented a number of measurement challenges. The documentation would require a level of accuracy and graphic representation that captured not only overall form, but the building's many details, such as the stone coursing and ornamentation. A final 3D model of processed and aligned scan data would serve as a basis for preparation of digital CAD drawings, in both 2D and 3D format.

7.6.3 The Scanning Process

A Faro X330 laser scanner was used to scan the building (Figure 7.14). Scanning began with a series of scans taken around the perimeter of the castle. Multiple target spheres were set around the facility to enable reference marks for the scans. Scans were taken at high resolution, providing a density level of scan points adequate to capture the details required by the client. Each location of the scanner was established to provide line of sight coverage of the building's shape and details. Overlap between adjacent scan

Figure 7.14 Use of an extension tripod allowed for better capture of data at upper levels

positions helped to add more detail and aid proper alignment. All scans were colorized to enable a final 3D image that captured the color and texture of the building.

As the exterior perimeter scans were nearing completion, the scanner was taken inside and brought up to the building's roof. From the parapet edge of the roof a scan was taken to capture the target spheres placed for the ground-level exterior scans. This established a reference basis to tie and align the roof scans to those around the perimeter. Additional scans were then taken to cover the various roof levels and complete the exterior scanning process. Some checkerboard targets were left in place near the doorway access to the roof; they will serve as a reference to the interior scans to be taken later.

The Castle had two main entry doors on opposite sides of the building. Before scanning the interior, a number of target spheres and checkerboard targets were placed in key locations throughout the main floor. The doorways were propped open, and a scan was taken from the doorsills that captured exterior targets left in place from the perimeter scanning along with the newly placed interior targets. These reference scans, used in conjunction with the rooftop reference scans, insured that the interior scan data could be aligned to the exterior scans. This is key to providing an accurate, complete 3D digital model of the castle.

With the exterior reference scans completed, the interior scanning process of the facility began. The main floor was scanned from a series of locations taken in a controlled pathway through the main level, ending at the main stairway. This became the stopping point for the first day's scanning efforts. Targets were left in place to allow a continuation of the control path during the next day's scanning.

The next day's scanning began at the entry to the cellar stair. Scans were taken at various locations going down the stair, being careful to maintain line of sight to all targets placed for reference. Once in the cellar, a similar process was followed; setting a control path with reference targets throughout the cellar and then scanning from multiple scanner locations.

After completing the cellar level, the scanner was taken back to the main level and placed at the base of the stairway to the upper levels. A reference scan was taken to capture existing targets in place at the main level to locate the scanner's starting position for the next series of scans. Similar to the cellar stair descent, targets were set going up the stairway, and scanning began. As each succeeding level was reached, more targets were placed enabling each floor to be scanned using a reference basis common to the entire building. When completed, over 110 scans had been taken to document the exterior and interior of the facility. The individual scans were ready for the next phase of the documentation process, to be downloaded, processed, and aligned into a complete 3D digital model.

7.6.4   Scan Registration

All scans were downloaded into Faro Scene, a proprietary software from the scanner's manufacturer that was developed to process and align scan data. The first stage of the software uses various algorithms within the software to filter each scan of noise and extraneous data. Following that process, the scans were aligned using the scanned targets as locational references for each scan. As the data was aligned, it was clustered into small subsets of scans. This helped to reduce the load on the computer's memory and processing speed. The final group of clusters, all aligned to the referenced targets, formed a complete 3D digital image of the building. This final set of data was then run though a cloud-to-cloud registration process within the Scene software to reduce overall scan drift and fine-tune the accuracy of the final digital model (Figures 7.15 and 7.16). Overall dimensional accuracy of the final model was +/−.75 inch, well within the project's parameters. The final scan files were exported into a 3D file format suitable for importation into CAD programs, ready to serve as a basis for 2D and 3D cad drawings (Figure 7.17).

Figure 7.15 Multiple scans combine to form point cloud view of building within scan processing software

Figure 7.16 Section view through building interior of point cloud data showing alignment of all levels

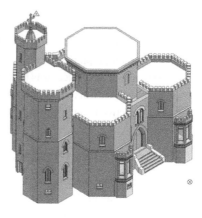

Figure 7.17 Point cloud image with resulting 3D CAD model

7.7 INFRASTRUCTURE PRESERVATION

The maintenance and repair of our country's infrastructure creates a need for an ongoing record of both completed and anticipated work items. 3D laser scanning's ability to quickly and accurately document existing conditions, yielding both visual and forensic information on surfaces, is becoming an important part of the workflow in developing infrastructure maintenance programs.

7.7.1 Project Scope

The Claiborne Pell Bridge connects Jamestown to Newport, RI. It is a suspension bridge, built in 1966, which covers a length of approximately 11,250 feet from abutment to abutment, with a suspension span of 1,608 feet. Bridges are especially susceptible to degradation over time. The salt air, constant traffic, and daily expansion and contraction of the bridge's structural components all contribute to breakdown of the steel supports and road surface. Well-planned maintenance and repair is critical to providing safe long-term operation of the bridge.

While undergoing a structural maintenance program, the Rhode Island Turnpike and Bridge Authority (RITBA) wanted to explore the possibility of setting up a database that would catalog and identify all the main parts of the upper roadway, i.e., railings, scuppers, curbs, light poles, and most importantly, the numerous areas of patched asphalt concrete roadway. The ability to document the road surface within a cad environment would allow them to understand the percentages of roadway currently patched, the age of those patches, and where future monies might be required for new or replacement patches. It would also provide a base model of the bridge that the RITBA could add to over time, resulting in a real-time model that updated modifications and repairs to both the road surface and structure.

Laser scanning was seen as a means to acquire the raw information needed for this database. The scan data had to be of high definition quality, so that individual scans could serve as photographic references as well as measurable data within CAD. The primary objective of the project was to document and draw the roadway with all patches and expansion joints displayed in scale. A method of scanning that would yield adequate data for that objective had to be proven before the project could proceed.

7.7.2 Scanning Process

Options of both mobile scanning and medium range terrestrial scanning were explored. Test scans were done off site to see the level of detail that could be expected from a road surface when scanned at night. The resultant data was also studied for its ease of use within CAD. Each scan had to be located and tagged to a roadway map, enabling a single scan to be quickly located and accessed on the computer for study. After considering these and other factors, including scanning speed of the laser scanner, station setup distances and targeting requirements, the decision was made to use a FARO X330 midrange scanner to scan the bridge. A test was conducted at the bridge; scans were taken at various levels of resolution and distances apart. The test data was also imported into CAD, to verify its suitability for producing a 2D CAD document that would be the final deliverable. Based on the test results, it was determined that scans set a distance of +/− 60′ (18 m) at a resolution of 1/4 would yield a point cloud data return best suited for the project's requirements.

Two scanning runs of 11,250 LF had to be completed within a timespan of ten working nights, over four lineal miles of road to scan. Access to the bridge was on weeknights only, from approximately 9:00 p.m. to 4:30 a.m. With setup and breakdown, the team had a working scan time of approximately 6 hours. The bridge is four lanes, split with spaced traffic dividers in the center, creating a two-lane roadway on each side. One side was shut down for traffic, allowing a clear pathway for scanning. The scanning team started at the beginning of the eastbound lane, and proceed scanning westward (Figure 7.18). Once that side was completed, the team began at the other end of the bridge and scanned back on the westbound lane. Scan locations were alternated on the west and eastbound sides, so final scan coverage of the bridge had scans placed roughly 30′ apart. This provided an overlap of data from lane to lane, as well as providing clear scan data for each of the eastbound and westbound traffic lanes (Figures 7.19 and 7.20). The scanning team included a survey crew who used a Leica Scan Station for survey control, capturing existing bridge benchmarks and all scanning targets. This allowed for controlled geodetic alignment once all data was processed.

Figure 7.18 Nighttime scanning on bridge using multiple scanners and total station

Figure 7.19 Adjusting scanner at fixed spacing intervals and heights to control scan density of road surface

Figure 7.20 View of scan data showing level of detail captured on road surface

7.7.3 Deliverable

A deliverable of the roadway surface and the two adjacent sides of the bridge illustrating the railings, suspension cabling, and towers was to be provided in CAD format. This CAD document would serve as the basis for indexing an excel database which would identify and describe the various items delineated.

When completed, 380 scans were taken on the bridge over a period of nine nights. Both FARO SCENE and Leica's Cyclone were used to process and align the data, with a final deliverable put together comprising all scans in Faro SCENE (Figure 7.21). All data was keyed to a scan location map that enabled quick identifying and viewing of an individual scan for a specific area of the bridge.

The final processed and aligned scan data was exported into Autodesk's ReCap, setup in a CAD file and folder system that mimicked the point cloud data. This avoided any confusion when searching for data from a specific scan location. The recap files were brought into AutoCAD, and used as a basis to create a plan view and elevations of the entire bridge. The data enabled location of light poles, railings, expansion joints, manholes, and scuppers, but most importantly the numerous areas of patches that were captured on the road surface (Figure 7.22). The success of the deliverable confirmed LIDAR's value as a database for visual offsite reference, and as a basis for deriving large-scale informative CAD files illustrating the forensic findings of point cloud data.

Figure 7.21 A view of multiple point cloud scans showing a portion of the bridge. Nighttime scanning provided excellent return of laser light points, yielding dense high-quality data

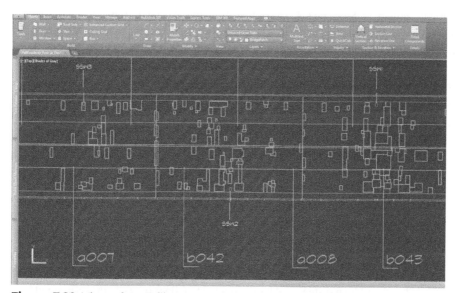

Figure 7.22 View of CAD file prepared indicating location and configuration of the patches on the road's surface

7.8 GENERATING STATION TURBINE REPLACEMENT

7.8.1 Project Scope

Brookfield Renewable Power's Bear Swamp Generating Station was developed to deliver supplemental electricity during periods of peak loads – periods when consumers place the heaviest demand on electricity. Completed in 1974, the Bear Swamp complex is a pumped-storage, hydroelectric underground power station that lies along the Deerfield River. With a natural reservoir atop a mountain and a river flowing at the base, the energy potential for this coincidence of nature was remarkable.

Deep within the mountain is Cockwell Station, a highly automated hydroelectric plant that uses a very unique method of generating electricity. During peak demand hours, the station operates as a conventional hydroelectric generating station. Released from the 118-acre upper reservoir, 10,760 cubic feet of water per second is channeled through a series of tunnels and vanes to drive two turbines and then flows into the lower reservoir. Operating in unison, the two turbines provide a total of 600,000 kilowatts of power that is delivered to the New England power grid.

At night, when energy demand drops, the turbines are reversed and become 415,000-horsepower motors that pump 8,800 cubic feet of water per second 770 feet from the lower reservoir back up into the upper reservoir until it is needed for the next peak period.

Bear Swamp needed to replace their current turbines and generators with higher capacity equipment. Since no CAD models or drawings existed for these turbines, laser scanning was selected as a means to measure the wicket gates and generate 3D models of the turbines in their as-built condition. These models could then be used in the design and construction of the newer turbines and generators, optimizing water flow and creating higher amounts of electricity in a clean, efficient manner.

7.8.2 Laser Scan Documentation Process

Due to the daily need for the facility's power, downtime was severely limited. It was absolutely critical that the needed measurements were taken efficiently, accurately, and quickly. In order to get the job done, a FARO Laser Tracker was used (Figure 7.23). The FARO Laser Tracker is an extremely accurate, portable contact measurement system that uses laser technology to perform measurements (Figure 7.24). The tracker uses a laser distance meter and two rotating axes to track the exact position of a mirrored spherical probe that is placed at numerous points along the object to be measured. The laser beam is reflected back to a position sensing device that then drives the servo motors in the laser tracker. This closed loop system enables the tracker to follow, or track, the reflector at 1,000 updates per second. The software then stores the measurement values, creating a database of 3D points describing the objects measured.

Figure 7.23 Obtaining measurements within the tunnel using a laser tracker and handheld CMR

Figure 7.24 Recording all measurements on a computer tablet connected to the laser tracker

7.8.3 Deliverable

The data, once complied and processed, provides a digital 3D database which can be used as a basis for generating a CAD model of the objects to be redesigned. The software can also compare the points to a prescribed tolerance range to better understand the condition of the existing equipment. With the accurate data provided by the FARO Laser Tracker, the Bear Swamp Power Company was able to bid out the contract for an upgraded turbine runner and generator. This new equipment increased Bear Swamp's capacity for low-cost power throughout New England. Normally a complex project of this scope would have been extremely difficult and time consuming. However, by using laser trackers that enabled high accuracy, efficient data capture the job was completed within Bear Swamp's scheduled downtime.

REFERENCES

East coast metrology: reverse engineering the brookfield renewable power plant. https://eastcoastmetrology.com/resources/applications/brookfield-renewable-power-plant. (accessed 18 March 2022).

McLaughlin, J. Tracking the Tokamak – the Holy Grail of renewable energy. https://eastcoastmetrology.com/resources/applications/tracking-the-tokamak/. (accessed 18 March 2022).

Roe, G. (2016). Case Study: fonthill Castle, John Smits. https://www.faro.com/en/Resource-Library/Case-Study/conserving-a-landmark-fonthill-castle. (accessed December 3, 2020).

Smits, J. (2014). Preserving infrastructure with point clouds. *LIDAR news*. (September 2014) 33.

Chapter 8
Future Directions

Today's advances in laser scanning provide the basis for capturing a digital image of an object, developing it within CAD software and printing a 3D replica at any scale (Figure 8.1). What does the future hold for scanning technology? How will it be used? What developments and refinements are anticipated in the next 25 years?

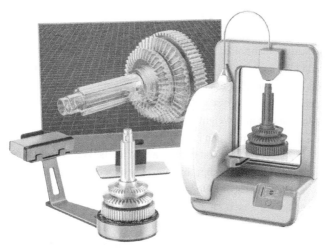

Figure 8.1 Scanning provides the basis of information allowing one to go from object to CAD file to 3D print

3D Scanning for Advanced Manufacturing, Design, and Construction, First Edition.
Gary C. Confalone, John Smits, and Thomas Kinnare
© 2023 John Wiley & Sons, Inc. Published 2023 by John Wiley & Sons, Inc.

8.1 SOFTWARE DEVELOPMENT

Along with the rise of cloud file storage, computer's exponential increase in levels of computing capacity has made possible development of advanced graphic software for 3D data. Many of these software are designed with particular user groups in mind, intended to serve the needs of their specific 3D applications. The growing market of specialized niche businesses within the greater 3D industry will drive even greater future development of software. Software as a Service (SaaS), the delivery of software through a cloud-based license system, will help to drive this trend. Its results will be twofold: as software allows greater use of 3D data, new technology devices to capture that data will be designed. These new devices in turn will allow further specialization of 3D niche industries, giving rise to more software development.

Software development will adapt in its usefulness as well. For example, until recently, software within the AEC industry has often been limited to a specific unique purpose, marketed to and serving a small segment of defined users. As software becomes more adaptable and widespread in its applications, it will allow for increased use of the new technological equipment. This in turn will provide for new and faster workflows which will drive the use of 3D laser scanning and its applications within the industry among all users.

8.2 TECHNOLOGY

One aspect of scanning technology is certain: there has been a consistent effort to make the scanning devices smaller, lightweight and thus increase the portability of the scanning device. Handheld devices typically rely on a connection to a laptop in order to run. To be truly independent, a scanning device must incorporate a laser, computer, storage, and battery in one lightweight portable device. Cell phones are the simplest of handheld scanning devices; most can run rudimentary forms of scanning applications. However, in lieu of a laser they rely on photogrammetry as their scanning method. That coupled with their lower level of computing power produces 3D scan data that is often far below the accuracy levels needed for engineering and design applications. The main issue to be resolved for increased capability and portability are designs that reduce size and weight while expanding capacity of these important scanner elements:

- Power Supply. Lasers need power to operate; the more powerful the laser the more power required. A portable unit must rely on battery power. Batteries add weight and size to a device. Battery technology will have to evolve to meet the demands of new smaller scanner designs.

- Computer Processing. Scanning units must have a built-in processor to run the laser and complex algorithms required for scanning. It also must be able to store the large quantities of scan data collected on internal RAM or a small removable SD card. For instantaneous mapping by vehicles or drones, solid-state silicon microchips are being developed for Lidar sensors that can count and store over one trillion photons per second.

- Laser or Structured Light Source. Unlike a simple handheld laser pointer, the laser beam in a scanner must be highly focused and directed. In addition to the light source, a mechanism (detector) must be in place to record the return of a beam, or a camera must be provided to record structured light deviations. These can add sizable weight and bulk to the equipment's design.

Lidar technology has a few limitations. It is an optical sensing system and can be affected by bright light or fog, environmental conditions that will disturb the laser pulse and its returns. As the aforementioned list illustrates, it also requires various other technologies to be integrated in order to have a viable piece of scanning equipment, including computers for data collection and compilation, transmitters for Bluetooth or cellular transmission of data, and a camera if color detection is needed. With this in mind, efforts are being concentrated on the LIDAR sensor, enabling it to capture a greater volume of point data faster while being reduced to as small a size as possible.

This has led to the development of LIDAR that utilizes solid-state sensors, allowing for a vast reduction in size over conventional lidar pucks. Flash Lidar is another technology that is rethinking traditional LIDAR sensing; it allows an image to be acquired by a single pulse, similar to a camera. These new sensor systems need to be further refined as they currently lack the distance detection and brightness of traditional LIDAR technology. Multiple companies are already investing R&D monies and integrating these new sensing technologies into specialized product applications. When perfected, these miniature sensors will collect in a nanosecond well defined, dense point cloud data.

Smaller scanning devices are usually limited in the distance they can scan as well as their level of precision. This is primarily due to their diminished laser power. The design of future scanners will involve trade-offs between scanning distance, accuracy, and unit size to provide equipment that can rapidly scan and process small parts or large areas while still providing a level of imagery and accuracy that is beneficial to the task at hand. Industrial facilities are already assimilating scanning devices into their quality control and manufacturing equipment. As the large-scale facility planning and real estate market grows, both static ground-based scanners and SLAM devices will evolve to serve these needs.

8.3 EXTENDED REALITY

As the evolution of the enterprise environment continues in the areas of corporate and facility infrastructure, Extended Reality (XR) technologies will become more developed and widely used. Extended reality is a generic term for various technologies that enhance our view of the world. They include Virtual Reality (VR), Augmented Reality (AR), and Mixed Reality (MR). Many of these technologies are made possible by the data provided by 3D laser scanning imagery (Figure 8.2).

Figure 8.2 Some of the many future applications of augmented reality include engineering design review, interior decorating, facility planning, equipment maintenance, and Building Information Modeling (BIM)

Virtual reality immerses the user in a computer-generated virtual environment. It is prevalent in entertainment and gaming, and is gaining use in the social media interactive environment. In order to meet this growing demand, capabilities such as volumetric data capture, motion tracking, and facial expression capturing will be needed, all of which have a basis in 3D scanning technology.

Augmented reality is a means of enhancing our view of the world by the overlaying of computer generated information. An example of this is the use of a smartphone capturing an image within an app that generates a text or spoken description of the item or area viewed. It can also provide visual information, such as an overlay of a 3D CAD drawing of an object, which can be employed when assembling or using a piece of equipment or machinery.

Scanning's use will also continue to grow within the XR industry as it provides a quick and inexpensive method to secure the information database needed to interface with VR and AR software. Smaller LIDAR units may even become part of future XR devices. Immersive VR goggles could one day have a lidar unit imbedded in them, allowing one to interact with the local built environment while streaming a digital image to be shared with others offsite. Combined with advances in software, these images could be rendered in real time to create seamless solid models for collaborative input across multiple disciplines. The potential for applications is endless, from allowing a complete immersion in an alternate reality within the comfort of your home to safely guiding one though a multistage hazardous industrial work process.

8.4   UAV AND AV

The UAV (unmanned aerial vehicle) industry has seen exponential increase in use over the past years, thanks to both aviation drone technology and improvements in LIDAR sensing. The growth of Autonomous Vehicles (AV) has also been aided by the development of sophisticated LIDAR sensing technologies in both hardware and software. This need for LIDAR in the UAV and AV industries has quickly become the next large market segment driving laser scan technology and sensing equipment. LIDAR will play a key role in the development of self-driving vehicles as its sensors are key to enabling a vehicle to locate itself in space among both static and moving objects. LIDAR outfitted drones facilitate large-scale mapping and multifaceted information collecting on a scale not possible with ground-based units, even surpassing mobile scanning. LIDAR will become our eyes within the machinery and equipment of the future, helping to increase our safety in a busy complex world (Figure 8.3).

8.5   INDUSTRY TRENDS

Key industries which will utilize more 3D scanning and data bases will be the areas of entertainment, sports, manufacturing, and AEC. Entertainment will benefit from growth of the use of computer-generated special effects and imagery. Scanning of actors is already commonplace in cinema and television production. Databases of artists and actors physical forms could revolutionize the ways shows are produced, providing a ready source of images accessible to small independent filmmakers.

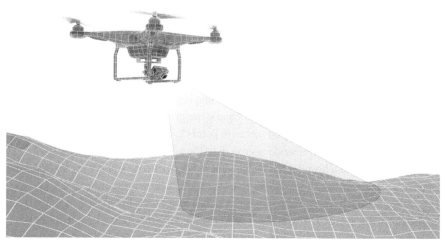

Figure 8.3 Drones incorporate LIDAR sensing to map terrain and objects

The sports industry will see scanning grow as a vehicle to assist in training. Whether scanning an athlete's movements while in action or scanning a race course, the resulting data can be used to analyze the best possible methods and approach to attain faster, error-free results during a game or race. Scanning's use in capturing as built imagery of stadium and venue seating will increase, allowing operators to design their seating, stage sets and playing field to maximize the line of sight view of attendees.

Manufacturing will see a surge in development of new designs, as it becomes easier to create a 3D image of an idea, make it on a 3D printer, test its effectiveness, and immediately begin redesign to fine-tune the product. Factories will employ LIDAR sensors to control movements of warehouse equipment, allowing for increased production capacity while saving their employees from the stress of repetitive tasks. Sensors within the facilities will accumulate data on everything from pathway travel densities to temperature fluctuations; all will be uploaded to a 3D BIM model that will be updated on a regular basis. The resultant data will provide for a safer and more energy efficient work environment.

The AEC industry has been trending toward increased use of visualization. The rise of remote workers, a highly developed internet and ability to cloud share large volumes of data has helped to drive this trend. In the years to come it will be commonplace for a worker to both see and measure what is happening at a site from the comfort of home or office, avoiding a day at a construction site in extreme heat or rain. In order to achieve that there will be a demand for multiple ways to collect and disseminate visual imagery tied to point cloud data, since "dumb" unscaled data has little use in construction and design. Present-day technology has already developed methods for capturing large amounts of data with little human interaction, such as the use of drones or mechanical doglike robots that can be programmed to scan a specific route within a building or construction site.

In the AEC and Facilities environment the trend is to merge the physical and digital worlds via augmented realty (AR). Digital twins of a large industrial space are created by laser scanning, which in turn can be visualized with AR. This dataset can be used for remote training, trouble shooting, and collaboration among departments and engineering disciplines.

Orthopedics are using magnetic resonance imaging (MRI) scans to create point clouds and produce 3D printed joints and bone structures. These custom joints and surgical inserts are made possible by advances in point cloud development and scan to print technology that have seen tremendous growth in the past decade (Figure 8.4).
Picture this: you go to a hospital for a knee joint replacement. As you are prepared for the operating room a technician waves a device over your knee. Within minutes a 3D printer has formed an exact duplicate of your joint which is ready to be installed. The scan data has also given the doctor a 3D visual of your joint and all surrounding ligaments and tendons, allowing him to plan his surgery down to the smallest detail. The reality of this scenario is not far off.

Similarly, metrology OEMs are developing units that will allow machined parts to be placed in a device that will quickly scan and create a 3D point cloud without contact-

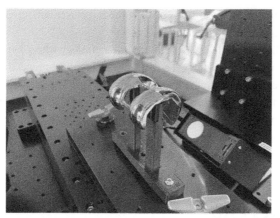

Figure 8.4 Prosthetic knee in 3D scanning device
Source: Photo Courtesy of DWFritz Automation, LLC.

ing the parts surface. The futuristic design of the *DW Fritz ZeroTouch* was developed in the early 2000s and it is a good example of future trends (Figure 8.5). Programmable scanning units like these allow parts to be loaded into the machine where an automated

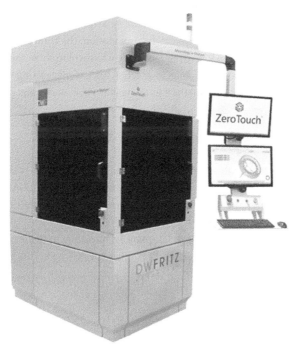

Figure 8.5 ZeroTouch automated 3D scanning device
Source: Photo courtesy of DWFritz Automation, LLC.

noncontact measurement process takes place without the user having to manually manipulate the part or measurement device.

Industrial applications will see more in-process metrology where parts can be compared to design in easy concurrent steps. The need for post-processing will disappear as this too will become automated. Companies are working toward increasing the potential of mechanized inspection processes which will allow quick and easy monitoring of parts as they are machined or developed. The future will see an increase in the use of 3D scanning technology integrated into the manufacturing process and production lines. This will provide instant results allowing for immediate corrections in the manufacturing process, whether it be minor adaptions to the process or a full work line stoppage.

8.6 SUMMARY

The obvious direction of the metrology and scanning industry is to develop instruments that are faster, lighter, and more accurate. Additionally the desire to do this without having to physically contact the surface of the object will become the ultimate goal. We are seeing recent trends that are moving the technology in this direction. The process of creating a model from a 3D point cloud will become a simple task in the future, as the digital imagery captured becomes a seamless combination of photos, xyz points, and solid raster CAD files, all driven by the initial laser scans. This collection of data will be available from a cloud-based internet, able to be viewed and used on multiple devices, from desktop computers to handheld tablets to personal smartphones.

In tandem with LIDAR's value in object documentation, the transportation industry will exploit the benefits of LIDAR sensing technology to revolutionize the way we move ourselves and goods from one place to another. 3D Laser scanning will be a basic building block of a transformative method in the way we live, work, and play.

REFERENCES

Arm Blueprint staff (2022). xR, AR, VR, MR: what's the difference in reality? http://www.Arm.com/blogs/blueprint/xr-ar-vr-mr-difference. (accessed 12 August 2022).

DWFritz Precision Automation. https://dwfritz.com/en/zerotouch-platforms. (accessed 4 January 2022).

Pacala, A. (2018). How multi beam flash lidar works. http://www.Ouster.com/blog/how-multi-beam-flash-lidar-works. (accessed 12 August 2022).

Chapter 9

Resources

The following are resources associated with the use, development, or promotion of 3D laser scanning and its applications in business and industry. They can assist in providing education and training, professional development and mentoring, and provide a source for certifications and standards.

Reference Books

- *Evaluation of Measurement Data – Guide to the Expression of Uncertainty (GUM)*, BIPM, 2008
- *International Vocabulary of Basic and General Terms in Metrology (VIM) Third Edition*, BIPM, 2006

Important Organizations

- American Institute of Architects (AIA)
- American Society of Mechanical Engineers (ASME)
- American Society for Nondestructive Testing (ASNDT)
- American Society for Photogrammetry and Remote Sensing (ASPRS)
- American Society for Quality (ASQ)
- American Society for Testing and Materials (ASTM)
- American National Standards Institute (ANSI)
- Coordinate Metrology Society (CMS)

3D Scanning for Advanced Manufacturing, Design, and Construction, First Edition.
Gary C. Confalone, John Smits, and Thomas Kinnare
© 2023 John Wiley & Sons, Inc. Published 2023 by John Wiley & Sons, Inc.

- International Bureau of Weights and Measures (BIPM) – French: Bureau International des Poids et Mesures
- International Laboratory Accreditation Cooperation (ILAC)
- International Organization for Standardization (ISO)
- International Organization for Standardization, Reference Materials Committee (ISO/REMCO)
- International Society of Automation (ISA)
- National Conference of Standards Laboratories International (NCSLI)
- National Institute of Standards and Technology (NIST)
- National Physical Laboratory (NPL)
- National Society of Professional Engineers (NSPE)
- National Society of Professional Land Surveyors (NSPS)
- Society of Automotive Engineers (SAE International)
- Society of Manufacturing Engineers (SME)
- The American Association for Laboratory Accreditation (A2LA)
- US Institute of Building Documentation (USIBD)

Hardware Manufacturers

- API — https://apimetrology.com
- Artec — https://www.artec.com
- Canon — https://www.usa.canon.com
- Carl Zeiss GOM Metrology GmbH — https://www.gom.com
- Creaform — https://www.creaform3d.com
- FARO — https://www.faro.com
- GeoSlam — https://geoslam.com
- Hexagon (Leica) — https://leica-geosystems.com
- Hokuyo-USA — https://www.hokuyo-usa.com
- Mantis — https://mantis-vision.com
- Matterport — https://matterport.com
- MetrologyWorks, Inc. — https://www.MetrologyWorks.com
- NavVis — https://www.navvis.com
- Nikon — https://www.nikonmetrology.com
- Ouster — https://www.ouster.com

- Riegl   https://www.riegl.com
- Surphaser   https://surphaser.com
- Topcon   https://www.topconpositioning.com
- Trimble   https://geospatial.trimble.com
- Velodyne Lidar Inc   https://velodynelidar.com
- Z$^{\&}$F   https://zr-usa.com

Software

- 3D Systems – Geomagic
- Autodesk Revit, Recap Software
- Bentley Software
- ClearEdge3D
- Cyclone – Leica Geosystems
- ESRI GIS Mapping Software
- FARO Scene Point Cloud Processing
- FARO CAM2 3D Measurement Software
- Hexagon – Inspire Software
- New River Kinematics - SpatialAnalyzer
- Oqton Software
- Pointfuse Point Cloud Software
- Innovmetric - Polyworks 3D Software
- Verisurf Software

Tradeshows and Conferences

- ABX Architecture Boston Expo
- AIA Conference on Architecture
- ASME Annual Meeting
- ASQ World Conference on Quality and Improvement
- CMSC Coordinate Metrology Society Conference
- Quality Show
- Geo Week – SPAR 3D, IMLF, AEC Next
- IMTS International Manufacturing Technology Show
- International LIDAR Mapping Forum
- NCSL International Symposium

- NSPE Professional Engineers Conference
- RAPID + TCT Conference 3D Printing & Additive Manufacturing
- SME Annual Conference and Expo
- USIBD Realty Capture
- User Conferences: ESRI, FARO, Hexagon, Riegl, Trimble

College and University Metrology/3D Scanning Study Programs

The following Colleges and Universities have 3D Laser Scanning Training, Programs, Centers, or Curriculum:

- Ball State University, Muncie, IN https://www.bsu.edu
- Cal Poly, Pomona, CA – Civil Engineering Program www.cpp.edu
- California Polytechnic State University, San Luis Obispo, CA https://ceenve.calpoly.edu
- Carnegie Mellon University, Pittsburgh PA

  School of Computer Science, Robotics Institute www.cs.cmu.edu
- Central Michigan University, Mount Pleasant MI

  Dept. of Engineering and Technology www.cmich.edu/academics/colleges/college-science-engineering
- Central Washington University, Ellensburg, WA

  Geological Sciences – wide range of TLS applications – https://geology.cwu.edu
- Clemson University, Charleston, SC

  Clemson University Restoration Institute http://Clemson.edu
- Columbia University, NY, NY

  Graduate School of Architecture, Planning and Preservation www.arch.columbia.edu
- The Idaho Virtualization Laboratory, https://www.isu.edu/imnh/idaho-virtualization-lab/the-lab/
- Iowa State University, Ames, IA

  Center for Industrial Research and Service, http://ciras.iastate.edu
- North Carolina State University, Raleigh, NC, www.ncsu.edu
- Purdue Engineering, West Lafayette, IN, https://engineering.purdue.edu

- Rochester Institute of Technology, Rochester, NY

  Imaging Science Degree, www.rit.edu

- Rensselaer, Troy, NY

  Center of Subsurface Sensing and Imaging Systems, www.rpi.edu

- School of the Museum of Fine Arts at Tufts University, Medford, MA https://smfa.tufts.edu

- Southern Illinois University, Carbondale, IL

  Environmental Resources and Policy, https://siu.edu

- Stanford University, Stanford, CA

  Stanford Archaeology Center – https://archaeology.stanford.edu

- Texas A&M University, College Station, TX

  School of Architecture, www.arch.tamu.edu

- University of Arizona, Tucson, AZ, http://www.arizona.edu/

- University of Arkansas – Center for Advanced Spatial Technologies, https://cast.uark.edu

- University of Connecticut, Storrs, CT

  Remote Sensing & Geospatial Data Analytics, https://uconn.edu

- University of Northern Colorado, Greely, CO, https://www.unco.edu

- University of South Florida, Tampa, FL

  Center for Digital Heritage and Geospatial Information, https://dhhc.lib.usf.edu/

- University of Texas, Dallas

  Cyber Mapping Lab, https://labs.utdallas.edu/cybermapping/home/ccli/

- The University of Texas at El Paso

  The Keck Center, https://utep.edu/keck

- University of Virginia, Charlottesville, VA

  The Institute for Advanced Technology in the Humanities, www.virginia.edu

- Virginia Tech, Blacksburg, VA

  Dept. of Forest Resources and Environmental Conservation, www.vt.edu

The following Colleges and Universities have Training, Programs, Centers, or Curriculum in Metrology and Quality Assurance:

- Alamo Colleges: St. Phillip's College, San Antonio, TX
  Precision Tools and Measurement
- Butler County Community College, Butler, PA
  Measurement Science/Metrology Technology
- Bridge Valley Community & Technical College, Charleston, WV
  Instrumentation, Measurement and Control Technology
- Broome Community College, Binghamton, NY
  Advanced Materials Manufacturing and Metrology Training Program
- California State Dominguez Hills, Carson, CA
  BS in Measurement Science
- Central Georgia Technical College, Macon, GA
  Metrology Associate Degree
- Danville Community College, Danville, VA
  Dimensional Inspection Metrology
- De Anza College, Cupertino, CA
  Quality Control Technician Certificate of Achievement
- Fox Valley Technical College, Appleton, WI
  Production Inspection & Metrology
- Fullerton College, Fullerton, CA
  Metrology Certificate of Achievement
- Fulton Montgomery Community College, Johnstown, NY
  Quality Control and Metrology Technician
- Goodwin College, East Hartford, CT
  Metrology
- Great Bay Community College, Rochester New Hampshire
  Advanced Technology Program
- Louisiana Delta Community College, Monroe, LA
  Industrial Instrumentation Technology
- Monroe County Community College, Monroe, MI
  Metrology Technology
- Nashua Community College, Nashua, NH
  Metrology and Quality Control for Precision Manufacturing Certificate

- Navajo Technical University, Crownpoint, NM

  Industrial Metrology Education
- Piedmont Technical College, Greenwood, SC

  Precision Metrology Certificate
- Polk State Corporate College, Winter Haven, FL

  Advanced Manufacturing Institute
- Red Rocks Community College

  Metrology Quality Control Certificate
- Tennessee Tech University, Cookeville, TN

  Graduate Studies in Mechanical Engineering – Metrology
- Tidewater Community College, Norfolk, VA

  Metrology Certificate
- University of North Carolina, Charlott,e NC

  William States Lee College of Engineering – Center for Precision Metrology
- Washington Polytechnic Institute, Bellingham, WA

  Manufacturing and Metrology Engineering
- Weber State University, Ogden, UT

  Certified Calibration Technician

WEBSITES

https://www.qualitydigest.com/
https://www.qualitymag.com
https://lidarnews.com/
https://www.ncsli.org/te/
https://www.cmsc.org/
https://metrology.news/
https://eastcoastmetrology.com/
https://laserfocusworld.com/
https://geospatialworld.net.com/

9.1   UNITS OF MEASURE

A. Measurement

Magnitudes of measurements are typically given in terms of a specific unit. In surveying, the most used units define quantities of length (or distance), area, volume, and horizontal or vertical angles. The two systems used for specifying units of measure are the English and metric systems. Units in the English system are historical units of measurement used in medieval England which evolved from the Anglo-Saxon and Roman systems. The metric system is a decimalized system of measurement developed in France in late eighteenth century. Since the metric system is almost universally used, it is often referred to as the International System of Units and abbreviated SI.

1. Length

a. English Units

The basic units for length or distance measurements in the English system are the inch, foot, yard, and mile. Other units of length also include the rod, furlong, and chain.

$$1\,foot = 12\,inches$$

$$1\,yard = 3\,feet$$

$$1\,rod = 5.5\,yards = 16.5\,feet$$

$$1\,chain = 4\,rods = 66\,feet = 100\,links$$

$$1\,furlong = 10\,chains = 40\,rods = 660\,feet$$

$$1\,mile = 8\,furlongs = 80\,chains = 320\,rods = 1{,}760\,yards = 5{,}280\,feet$$

b. Metric Units

The basic unit of length in the SI system is the meter. The meter was originally intended to be one ten-millionth of the distance from the Equator to the North Pole (at sea level). The meter has since been redefined as the distance traveled by light in a vacuum in 1/299,792,458 seconds (i.e. the speed of light in a vacuum is 299,792,458 m/sec). Subdivisions of the meter are the millimeter, centimeter, and the decimeter, while multiples of meters include the decameter, hectometer, and kilometer.

$$1\,meter = 1{,}000\,millimeters$$

$$1\,meter = 100\,centimeters$$

$$1\,meter = 10\,decimeters$$

$$1\,decameter = 10\,meters$$

$$1\,hectometer = 100\,meters$$

$$1\,kilometer = 1{,}000\,meters$$

c. English to Metric Conversions

There are two different conversions to relate the foot and the meter. In 1893, the United States officially defined a meter as 39.37 inches. Under this standard, the foot was equal to 12/39.37 m (approximately 0.3048 m). In 1959, a new standard was adopted that defined an inch equal to 2.54 cm. Under this standard, the foot was equal to exactly 0.3048 m. The older standard is now referred to as the US survey foot, while the new standard is referred to as the international foot.

$$1\,meter = 39.37\,inches$$

$$1\,meter \times \dfrac{39.37}{12} \cong 3.2808\,feet$$

$$1\,foot \times \dfrac{12}{39.37} \cong 0.3048\,meter$$

$$1\,mile \cong 1,609.4\,meters \cong 1.6094\,kilometers$$

2. Area

a. English Units

In the English system, areas are typically given in square feet or square yards. For larger area measurements, the acre or square mile may be used. Historically, the acre was originally established as an area one furlong in length and four rods in width. Laying out ten of these acres side by side is a square furlong (10 acres). Since a mile is eight furlongs in length, there are exactly 640 acres in a square mile. A survey township is a square unit of land six miles on a side that conforms to meridians and parallels. Each township is further divided into 36 one-square mile sections. Because some of the townships have boundaries designed to correct for the convergence of meridian lines, not all townships and their sections are exactly square.

$$1\,square\,foot = 12\,inches \times 12\,inches = 144\,square\,inches$$

$$1\,square\,yard = 3\,feet \times 3\,feet = 9\,square\,feet$$

$$1\,square\,rod = 16.5\,feet \times 16.5\,feet = 272.25\,square\,feet$$

$$1\,square\,chain = 66\,feet \times 66\,feet = 4,356\,square\,feet$$

$$1\,square\,furlong = 660\,feet \times 660\,feet = 435,600\,square\,feet$$

$$1\,acre = 4,840\,square\,yards = 43,560\,square\,feet$$

$$1\,acre = 1/10\,square\,furlong = 10\,square\,chains = 160\,square\,rods$$

$$1\,square\,mile = 1\,section = 640\,acres$$

$$1\,township = 36\,sections = 36\,square\,miles$$

b. Metric Units

Areas in the metric system are given in square meters while larger measurements are given in hectares.

$$1\,square\,meter = 1,000\,mm \times 1,000\,mm = 1,000,000\,square\,mm$$

$$1\,square\,meter = 100\,cm \times 100\,cm = 10,000\,square\,cm$$

$$1\,hectare = 100\,meters \times 100\,meters = 10,000\,square\,meters$$

$$1\,square\,kilometer = 1,000\,m \times 1,000\,m = 1,000,000\,square\,m$$

$$1\,square\,kilometer = 100\,hectares$$

c. English to Metric Conversions

$$1\,square\,meter \cong 1.1960\,square\,yards$$

$$1\,square\,meter \cong 10.7639\,square\,feet$$

$$1\,hectare \cong 2.4710\,acres$$

$$1\,square\,kilometer \cong 247.1044\,acres$$

$$1\,square\,mile \cong 2.5900\,square\,kilometers \cong 258.9998\,hectares$$

3. Volume

a. English Units

Volumes in the English system are typically given in cubic feet or cubic yards. For larger volumes, such as the quantity of water in a reservoir, the acre-foot unit is used. It is equivalent to the area of an acre having a depth of 1 foot.

$$1\,cubic\,foot = 12\,inches \times 12\,inches \times 12\,inches = 1,728\,cubic\,inches$$

$$1\,cubic\,yard = 3\,feet \times 3\,feet \times 3\,feet = 27\,cubic\,feet$$

$$1\,acre\;foot = 43,560\,square\,feet \times 1\,foot = 43,560\,cubic\,feet$$

b. Metric Units

Volumes in the metric system are given in cubic meters.

$$1\,cubic\,meter = 1,000\,mm \times 1,000\,mm \times 1,000\,mm = 1,000,000,000\,cubic\,mm$$

$$1\,cubic\,meter = 100\,cm \times 100\,cm \times 100\,cm = 1,000,000\,cubic\,cm$$

c. English to Metric Conversions

$$1\,cubic\,meter \cong 1.3079\,cubic\,yards$$

$$1\,cubic\,meter \cong 35.3145\,cubic\,feet$$

4. Mass

The mass of an object is often referred to as its weight though these are different concepts and quantities. Mass refers to the amount of matter in an object, whereas weight refers to the force experienced by an object due to gravity. In other words, an object with a specific mass will weigh more on the earth than the moon.

a. English Units

The avoirdupois pound is the primary unit of mass in the English system. Avoirdupois is a system of weight based on the 16 ounces per pound rather than the 12 ounces per pound in the troy system of weight.

$$1\,ounce = 16\,drams$$

$$1\,pound = 16\,ounces$$

$$1\,hundredweight\,(short) = 100\,pounds$$

$$1\,hundredweight\,(long) = 112\,pounds$$

$$1\,ton\,(short) = 2,000\,pounds$$

$$1\,ton\,(long) = 2,240\,pounds$$

b. Metric Units

The kilogram is the unit of mass in the metric system.

$$1\,gram = 1,000\,milligrams = 100\,centigrams = 10\,decigrams$$

$$1\,kilogram = 10\,hectograms = 100\,decagrams = 1,000\,grams$$

$$1\,metric\,ton = 1,000\,kilograms$$

c. English to Metric Conversions

The avoirdupois pound is defined as exactly 0.45359237 kg.

$$1\,pound \cong 0.4536\,kilograms$$

$$1\,kilogram \cong 2.2046\,pounds$$

$$1\,metric\,ton \cong 1.1023\,tons\,(short)$$

5. Angular Measurement

In geometry, any horizontal or vertical angle is measured in degrees. These angles may be given in decimal degrees or degrees, minutes, and seconds.

$$1\,degree = 60\,minutes = 3,600\,seconds$$

$$e.g.\,45.55\overline{5}\,degrees = 45°33'20''$$

The radian is another unit of measure for angles. By definition, a full circle has 2π radians or 360 degrees.

$$2\pi\,radians = 360\,degrees$$

$$1\,radian = \frac{360}{2\pi}\,degrees \cong 57.2958\,degrees$$

6. Temperature

a. English Units

The Fahrenheit scale, or degrees Fahrenheit (°F), is used in the United States to measure temperature. On the Fahrenheit scale, the freezing point of water is 32°F while the boiling point is 212°F at standard atmospheric pressure. The boiling and freezing points of water are exactly 180 degrees apart, making each degree Fahrenheit 1/180 of the interval between the two points.

b. Metric Units

The Celsius scale, or degrees Celsius (°C), is used as the metric system to measure temperature. On the Celsius scale, the freezing point of water is 0°C while the boiling point is 100°C at standard atmospheric pressure. The boiling and freezing points of water are exactly 100 degrees apart, making each degree Celsius 1/100 of the interval between the two points. The Fahrenheit and Celsius scales converge at −40° (i.e. −40°F and −40°C are the same temperature).

$$(212-32)^\circ F=(100-0)^\circ c; 180^\circ F=100^\circ C$$

$$1^\circ F=\frac{100^\circ}{180} C=\frac{5^\circ}{9} C$$

$$1^\circ C=\frac{180^\circ}{100} F=\frac{9^\circ}{5} F$$

c. English to Metric Conversions

$$^\circ F=\frac{9}{5}\times^\circ C+32$$

$$^\circ C=\frac{5}{9}\times(^\circ F+32)$$

7. Pressure

Atmospheric pressure is the force per unit area exerted against a surface by the weight of the Earth's atmosphere above that surface. Because there is less overlying atmospheric mass as elevation increases, pressure decreases with increasing elevation. The standard atmosphere (atm) is an international reference for pressure.

a. English Units

In the English system, air pressure is typically measured in inches mercury (inHg).

1 atm = 29.2125 $inHg$

b. Metric Units

Air pressure is measured in millimeters mercury (mmHg) or millibars (mbars) in the metric system but may also be measured in pascals or kilopascals.

$$1 atm=101,325 Pa=1013.25 mbars=760\ mmHg$$

c. **English to Metric Conversions**

$$1\,inHg \times \frac{1,000}{39.37} = 25.40\,mmHg$$

(WYDOT, 2011)

9.2 COORDINATE METROLOGY SOCIETY (CMS) CERTIFICATION

9.2.1 Certification Overview

Portable 3D Metrology is being integrated into manufacturing processes at a rapid rate. Data collection technologies that were once the domain of scientists, engineers, and mathematicians are now being used by technicians and shop-floor personnel for industrial applications and beyond. Measurement equipment is calibrated and certified to performance standards, but most of the personnel operating this equipment are not accredited. There are many variables induced by an operator that can dramatically influence data collection and dimensional control. It is important for an employer to verify the knowledge and skill level of an employee, or a metrology service provider in this industry.

The Coordinate Metrology Society determined a clear need for a Certification program based on a graying workforce, new skill requirements from industry, and the ever-increasing need for metrology expertise. Today, CMS Certification credentials aid in quantifying a potential employee's knowledge of metrology, which is essential to ISO certified manufacturers and companies with Quality Management Systems.

9.3 CERTIFICATION COMMITTEE

The CMS Certification Committee was formed in 2009 to assess the need and feasibility of creating a certification program for portable metrology professionals. An established charter was developed that defined the committee membership, the reporting responsibility, and updated the CMS bylaws to make certification committee a standing committee. The group researched existing certifications, surveyed the CMS membership (2009 conference), and began developing a preliminary Body of Knowledge (BOK). The group also determined the role of the Coordinate Metrology Society as an issuing authority. The bylaws governing the committee can be found online at www.CMSC.org/certification-committee.

After nearly five years of steady development, the Coordinate Metrology Society launched the industry's first Level-One Certification for 3D Portable and CMM Metrology. The first examinations were held at the 2013 CMSC in San Diego, CA.

The dedicated CMS Certification Committee is composed of distinguished metrology veterans and manufacturing industry professionals from around the globe.

9.3.1 Development of the Certification Program

At the onset, the CMS certification committee was formed to study the need for 3D Portable and CMM Metrology Certification. The original assumptions were to establish that equipment operators and data processors were the target audience. The committee investigated a partnership in administering certification and determined training would be provided by third-party organizations such as manufacturers, service providers, academia, and national institutes. Conclusions drawn from this 2009 study were:

- A properly structured certification program would be of definite value.

- Equipment must be calibrated, but the operator, the greatest potential source of error, is not required to be certified.

- Certification should be multilevel to delineate degree of capability and responsibility.

- Certification should indicate mastery of a core body of knowledge with additional certifications for equipment/software.

- Hardware/software certification should demonstrate appropriate technical knowledge and proficiency.

- There should be certified examiners for each hardware group.

- There may be areas where certification would be application specific.

9.3.1.1 CMSC 2010 Skills Development Workshop

The CMS Certification Committee developed the idea to perform a statistical study at CMSC. The aim of the study would identify skill gaps in the general metrology community. In addition to this, the workshops at CMSC would relate content to data developed in the study. The CMSC 2010 study was an open workshop inviting conference delegates to participate in an official Measurement Study. The core subject of the measurement study was based on a variety of "hand tools" used in dimensional measurement. The study's objectives were to look at the importance of core measurement principles, observing behavior when dealing with measurements, instilling the right measurement strategy, and encouraging a questioning culture.

The measurement studies were conducted over two days in two separate areas using various defined first principle tasks. The criteria of the tasks were modified to allow for various training and assessment techniques to be undertaken such as:

- Assessment of prior learning and experience

- Questioning techniques

- Practical task monitoring

- Demonstration

2011–2014 Measurement Studies

Since 2011, the Measurement Study theme has varied:

1. How Behavior Impacts Your Measurement in 2011

2. The Importance of Practical Testing in 2012

The annual Measurement Studies have greatly contributed to the body of knowledge used to produce the existing certification examinations. The 2012 measurement study was developed to support the organization's Certification Cognitive Examination development process, which culminated in the industry's first Level-One personnel certification program for portable 3D metrology and CMM metrology. The 2013 and 2014 studies were focused on the operator experience using 3D metrology devices.

9.4　CMS CERTIFICATION LEVELS

9.4.1　CMS Level-One Certification

The Level-One examination is a proctored and secure online cognitive assessment consisting of about 200 multiple-choice questions covering foundational theory and practice common to most portable 3D Metrology devices and CMM devices. Both of these exams are separate from one another, allowing candidates to choose the area of focus. Candidates for these certification must submit an application, meet eligibility requirements, sign the CMS code of ethics, and pass a peer review. Qualifying candidates are notified and scheduled for an examination seat. Assessment test questions will measure knowledge of metrology fundamentals identified in five job-duty areas as illustrated in Table 9.1.

Below are the basic requirements for the certification.

Table 9.1 Five focus areas of Level-One certification along with weighting percentage

Focus Area	Percentage of questions
Interpret design documents/ requirements	16%
Measurement device knowledge	16%
Premeasurement planning	23%
Performing measurement operations	25%
Analyzing data	19%
Ethics and professional responsibility	1%

9.4.1.1 Professional Certification Requirements

In view of the wide variety of skills and disciplines studied and practiced by portable 3D metrologists, the CMS Certification Program is based primarily on evidence of demonstrated professional capability. Within prescribed limitations, experience may be substituted for education and vice versa, and requires a written examination.

9.4.1.2 Basic Requirements

The basic requirements for a Level-One Certified Metrologist are:

1. A professional who uses portable or CMM 3D metrology technology to collect measurements and interpret data from a variety of contact and noncontact portable 3D measurement devices. The metrologist is responsible for all phases of data collection and other requirements such as planning and supervising survey activities requirements and interpretation.

2. Two years of experience in either portable 3D metrology or CMM metrology, one year of which was in a position of responsibility demonstrating professional knowledge and competence.

3. References from three people who are holding, or who have held, responsible positions in 3D metrology and have first-hand knowledge of the applicant's professional and personal qualifications.

4. Declaration of compliance with the Code of Ethics of the Coordinate Metrology Society.

5. Successful completion of the online examination: http://www.cmsc.org/cms-certification.

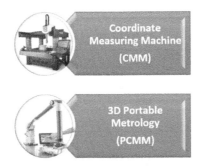 

Figure 9.1 There are two versions of the Level-One certification – 3D portable certification and 3D CMM certification. Both are online cognitive exams
Source: Image courtesy of Hexagon Metrology; *Reategui12 / Wikimedia Commons / CC BY-SA 3.0.*

9.4.1.3   Application for CMS Level-One Certification

To apply for the CMS Level-One Certification, you must complete a PDF application and submit Reference Forms, which can be downloaded from the CMS website: www.CMSC.org/cms-certification.

9.4.1.4   Preparation for the Level-One Certification Examination

In preparation for the exam, candidates should review the Examination Matrix which outlines appropriate paths of study. Knowledge in each job-duty task and 2 years minimum of hands-on experience with either portable or CMM 3D measurement systems will prepare you for this examination.

Preparation courses are available by approved proctor sites. Partnered with the National Physical Laboratory, *ECM-Global Measurement Solutions* created a Certification Preparatory course:

CMS Certification Preparatory Course: Level-One

A 2.5 day course designed to review the five areas of knowledge in the (CMS) Coordinate Measurement Society's Level-One Certification for 3D Metrology.

For information on this course visit ECM's website: www.eastcoastmetrology.com

9.5   CMS LEVEL-TWO CERTIFICATION

The Level-Two Certification examination for 3D laser scanning is a practical performance assessment. The candidate will use a laser scanner to collect a series of measurements on an artifact and then analyze specific features of that artifact.

Applicants for the Level-Two Certification must submit an application, meet eligibility requirements, sign the CMS Code of Ethics, and pass a peer review.

9.5.1   Professional Certification Requirements

In view of the wide variety of skills and disciplines studied and practiced by portable 3D metrologists, the CMS Certification Program is based primarily on evidence of demonstrated professional capability. Within prescribed limitations, experience may be substituted for education and vice versa, and requires a practical (hands-on) examination.

9.5.2   Basic Requirements

The basic requirements for a Level-Two Certified Portable 3D Metrologist are:

1. Must hold a current Level-One Certification
2. Two years of hands-on practical experience and a minimum of 400 hundred hours use, with the requisite device

3. Two references

4. Successful completion of a practical (hands-on) examination.

Note: Level-Two Certifications expire on same date as Level-One Certifications.

9.5.3   Application for CMS Level-Two Certification

To apply for the CMS Level-Two Certification, you must complete a PDF Application and submit Reference Forms, which can be downloaded from the CMS website: http://www. CMSC.org/cms-certification.

9.5.4   Preparation for the Level-Two Certification Examination

In preparation for this hands-on practical assessment, candidates should review the Assessment Matrix, which outlines seven performance areas evaluated during the Level-Two Certification test. Knowledge in each job-duty task and 2 years minimum of hands-on experience with portable 3D measurement systems will prepare you for this examination (Figure 9.2).

Currently there are four Level-Two Exams available for hands-on certification testing. These hands-on exams are available to be taken at approved proctor sites. Cur-

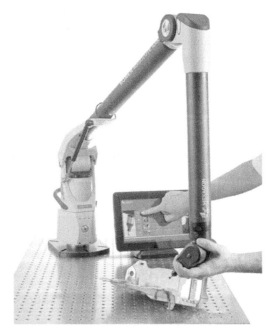

Figure 9.2 There are four versions of the Level-Two certification to date including 3D scanning and laser tracking. Tests are performed on an artifact focusing on ability and knowledge of instrumentation
Source: Image courtesy of Hexagon Metrology.

Figure 9.3 Preparatory courses and manuals are available through ECM –
Global Measurement Solutions, an approved CMS proctor site

rently *ECM – Global Measurement Solutions* is the approved Level-One and Level-Two
certification proctor site offering certification testing in the following specialties:

- 3D Laser Scanner
- Laser Tracker
- Articulating Arm
- Coordinate Measuring Machine

For information on testing preparatory training, visit ECM's Website:
www.eastcoastmetrology.com

9.6 NATIONAL METROLOGY LABORATORIES AND INSTITUTES

A National Metrology Lab or Institute is responsible for maintaining the national
measurement standards and infrastructure for the country for which it stands.

These NMIs represent an international network of laboratories that also provide
measurement and calibration testing facilities. Below is a list of a handful of NMIs that
may be a useful research tool for the reader.

Brazil

Instituto Nacional de Metrologia, Normalização e Qualidade Industrial (INMETRO)
Santa Alexandrina, 416–5 andar

Rio Comprido
20261–232, Rio de Janeiro
Phone: (55 21) 2679 9001
Fax: (55 21) 2563 1009
E-mail: homepage@inmetro.gov.br

Canada
National Research Council Canada
Institute for National Measurement Standards
1200 Montreal Road, Building M-36
Ottawa, Ontario K1A 0R6
Phone: (613) 998 7018
Fax: (613) 954 1473
E-mail: georgette.macdonald@nrc-cnrc-gc-ca

France
Laboratoire National d'Essais
LNE-Paris (hdqtrs) 1,
rue Gaston Boissier 75724 Paris Cedex 15
Phone: (33 1) 40 43 37 00
Fax: (33 1) 40 43 37 37
E-mail: info@lne.fr

Germany
Physikalisch-Technische Bundesanstalt (PTB)
Bundesallee 100
Postfach 3345
D-3300 Braunschweig
Phone: (49 531) 592 3006
Fax: (49 531) 592 3008
E-mail: info@ptb.de
Federal Institute for Materials Research and Testing (BAM)
Unter den Eichen 87
12205 Berlin
Phone: (49 30) 8104 0
Fax: (49 30) 8112 029
E-mail: info@bam.de

Japan
National Institute of Advanced Industrial Science and Technology (AIST)
1-3-1, Kasumigaseki
Chiyoda-ku, Tokyo 305–8561
Phone: (81 3) 55010900
India
National Physical Laboratory (NPL)
Dr. K.S. Krishnan Road

New Dehli, 110012
Phone: (91 11) 2574 2610
Fax: (91 11) 2572 6938
E-mail: root@nplindia.org

Ireland
National Metrology Lab (NML)
Forbairt
Glasnevin, Dublin 9
Phone: (353 1) 808 2609
Fax : (353 1) 808 2020
E-mail: nml@enterprise-ireland.com
National Standards Authority of Ireland
Glasnevin, Dublin 9
Phone: (353 1) 807 3800
Fax: (353 1) 807 3838
E-mail: nsai@nsai.ie

Israel
National Physical Laboratory
Danziger A. Building
Hebrew University
Jerusalem 91904

Italy
Istituto Nazionale di Ricerca Metrologica
Strada delle Cacce, 91 – 10135 Torino
Phone: (39 011 3919 1)
Fax: (39 011 346384)
E-mail: inrim@inrim.it
Istituto Nazionale di Metrologia delle Radiaziono Ionizzanti Roma (ENEA)
CR Casaccia c.p. 2400
Via Anguillarese 301
00060 Santa Maria Galeria RM
Phone: (39 6) 3048 3555
Fax: (39 6) 3048 3558
Consiglio Nazionale delle Ricerche (CNR)
Piazzale Aldo Moro, 7
00185, Roma
Phone: (39 06) 49931
Fax: (39 06) 4461954

Mexico
Centro Nacional de Metrología (CENAM)
km 4,5 Carretera a Los Cués

Apdo Postal 1–100
Centro, Querétaro 76900
Phone: (52 442) 211 0510
Fax: (52 442) 215 5332
E-mail: servtec@cenam.mx

Poland
Central Office of Measures
Ul. Elektoralna 2
00–139, Warsaw
Phone: (48 22) 620 02 41
Fax: (48 22) 620 83 78
E-mail: gum@gum.gov.pl

Sweden
Swedish National Testing and Research Institute
Box 857
SE-50115 15 Borås
Phone: (46 33) 16 5000
Fax: (46 33) 13 5502
E-mail: info@sp.se

Switzerland
Swiss Federal Office of Metrology and Accreditation
Lindenweg 50
CH-3003 Bern-Wabern
Phone: (41 31) 323 3111
Fax: (41 31) 323 3210
E-mail: info@metas.ch

United Kingdom
National Physical Laboratory (NPL)
Queens Road
Teddington, Middlesex TW 11 OLW
Phone: (44 181) 977 3222
Fax: (44 181) 943 6458
E-mail: enquiry@npl.co.uk

United States
National Institute of Standards and Technology (NIST)
100 Bureau Drive
Gaithersburg, MD 20899
Phone: 301 975–2000
E-mail: own@nist.gov
Full listing of NMLs:
https://www.nist.gov/iaao/national-metrology-laboratories

Metrology Glossary

The following is a glossary consisting of common industry terminology used within the metrology community. This collection of words, phrases, and abbreviations will assist in your understanding of the book's content as well as help you effectively communicate the concepts and workflows presented.

A

Accuracy The closeness of a measured quantity to the actual quantity that was measured, the measurand. Sometimes thought of as a tolerance range on a measurement's value. Often used colloquially as a synonym for uncertainty of measurement.

Alignment The state or orientation of an object or feature with respect to a set of datums or the act of putting an object or feature into a desired state or orientation. Example alignments include setting several bearings supporting a shaft in a line (or not quite a line); mounting aircraft sensors in a specific direction with respect to an aircraft's navigation system; boresighting weapons mounted on aircraft; best-fitting point cloud data to a set of surfaces.

As-built Or the As-built condition. Refers to the actual dimensions of a part or assembly as it was manufactured. Also sometimes called "As-found." In reverse engineering, as-built models or drawings do not try to divine design intent. They try to exactly portray a part exactly as it is, including defects.

ASCII (American Standard Code for Information Interchange) ASCII is an abbreviation for American Standard Code for Information Interchange, a standard data-encoding format for electronic communication between computers. ASCII assigns

3D Scanning for Advanced Manufacturing, Design, and Construction, First Edition.
Gary C. Confalone, John Smits, and Thomas Kinnare
© 2023 John Wiley & Sons, Inc. Published 2023 by John Wiley & Sons, Inc.

standard numeric values to letters, numerals, punctuation marks, and other characters used in computers. It is common for ASCII-formatted files to be used in transferring data between different software programs.

B

Ball-Bar A precision fixture that can be used as a standard of length. Ball-bars are often built with precision nests to accept Sphere Mounted Reflectors or precision ball bearings to aid in checking or calibrating laser trackers and CMMs.

Basic Dimensions Basic dimensions are the values on an engineering drawing that represent the theoretically exact size, profile, orientation, or location of a feature or datum target. Basic dimensions are identified by the dimensions on a drawing that are contained in a box drawn with a thin solid line.

Boresighting This term refers to an old method for setting sights on a firearm, in which the bolt or part of the action was removed, and a target point was viewed alternately through the bore and through the sights while adjustments were made until the sights pointed at the target point.

The term "Boresighting" today has been expanded to include determining the primary axis of a directional antenna, and aircraft weapons harmonization. Many techniques now exist including the use bore mounted lasers or laser trackers.

Bridge CMM Bridge-style CMMs provide probe movement along the three orthogonal axes (x,y,z). Each axis is monitored with extreme accuracy to define the center of the probe location in a 3D Cartesian Coordinate System.

C

CAD (Computer-Aided Design) CAD is the use of computer-based software to aid in the creation, analysis, or modification process of a design. CAD software is used by different types of engineers, designers, and architects to increase the productivity and overall product quality. CAD software can be used to create two-dimensional (2D) drawings or three-dimensional (3D) models.

Calibration "Calibration is the operation that, under specified conditions, establishes a relationship between the quantity values with measurement uncertainties provided by measurement standards and corresponding indications with measurement uncertainties." (VIM)

Thus, calibration involves measurements and comparisons, not an actual adjustment. For metrology, the formal comparison is of measuring equipment against a standard of higher level (a national standard defined in the US by NIST) under controlled and specified conditions to document the accuracy of the instrument being compared.

Control Point A measured point used in a set of at least three points to define a coordinate system. Each point is measured as the basis for improving the spatial accuracy of all other points and features to which they are connected through a least-squares adjustment. Control points are often used to relocate an instrument, transform coordinate systems, or to scale a subject that may be experiencing thermal change. Least-squares adjustments use control points to compute a solution by finding a minimum for the sum of the squares of the measurement residuals. Thus, a coordinate system can be defined using control points alone. The minimum number of control points needed to define a coordinate system is three.

Comma-Separated Values (CSV) A comma-separated values (CSV) file is a delimited text file that uses a comma to separate each value. Each line of the file is a data record. Each record consists of one or more fields, separated by commas.

In typical point cloud data files, each line of the file represents a point whose coordinates are stored as plain text in three comma separated fields for x,y,z. Other formats that store more data about each point are possible, for example each line could represent point coordinates and the direction of measurement using six fields of comma-separated text for x, y, z, i, j, k data.

Compensation Compensation is a technique or process used to compensate for error which may be inherently added during measurement.

Coordinate Measuring Machine (CMM) This acronym stands for Coordinate Measuring Machine. These machines measure parts through the acquisition of points.

Many types of CMMs exist; traditional 3 axis contact probing machines, articulated arm CMMs, and laser trackers are a few examples. Laser scanners or "white light" scanners are other examples that don't require contact with the measured part. CMMs are often run with software that can compute features such as planes, cylinders, and lines from the measured points.

Correction Correction is the quantity, in a measurement model, compensating for an estimated systematic error.The compensation can take different forms, such as an added or a factor, or can be deduced from a table.

CTE (Coefficient of Thermal Expansion) Coefficient of thermal expansion (CTE) is the ratio of the change in size of a material per degree change in temperature at a constant pressure.

D

Datum In metrology, a datum is simply an idealized reference feature from which another features' orientation, position, or other characteristic is defined. Datums can be many types of features, planes, cylinders, points, centerlines, constructions, or offsets from other features, anything that can be measured or established for use in locating other features.

Although the plural form of datum is really data, many people, including the author, uses datums as the plural in this application.

Datum Feature A datum feature is the measured or contacted feature which establishes a datum. It is an actual point, line, or surface on the part being measured.

Dimensional Inspection The process of characterizing an object's size and shape through measurements of points, lengths, and volumes. Usually such an inspection results in a report comparing the measured object to another object, an idealized object, or a previous state of the object.

Dimensional Metrology The study or practice of high-precision measurements to quantify physical sizes, orientations, and distances of objects and shapes.

Discrimination Threshold "The largest change in a value of a quantity being measured that causes no detectable change in the corresponding indication. NOTE: Discrimination threshold may depend on, e.g. noise (internal or external) or friction. It can also depend on the value of the quantity being measured and how the change is applied." (VIM)

F

First Article Inspection (FAI) As the name implies, first article inspection is usually performed on one or more of the first parts in production. The inspection is usually exhaustive, covering every dimension on the drawing, and provides a final verification of the manufacturing process. After the first article is proved out, subsequent parts are often inspected less exhaustively to save time and money.

Free-Form Shapes In metrology, this refers to surfaces with unconventional or continuously varying shapes like bones, customized molds, boat hulls, or the sculptures of Henry Moore.

G

Gage R&R (Gage Repeatability and Reproducibility) A set of repeated measurements used to determine the fitness of a gage or other measuring instrument for a specific function. The test tries to account for the effects of equipment, method, and operator in an estimate of precision and uncertainty by having a number of operators use the equipment to measure a reference standard or part.

Gantry CMM A gantry CMM has the measurement head mounted on a beam that overhangs and spans the measurement table. The head can move both vertically and side to side along the beam, providing two axes of motion. The third axis can be generated by moving either the towers that support the beam along a fixed table or moving the measurement table between the towers.

GD&T (Geometric Dimensioning and Tolerancing) Geometric dimensioning and tolerancing is a system and symbolic language for describing the permissible limits, or tolerances, in a part's physical dimensions and measured values. In the United States the standard describing GD&T is ASME Y14.5. In Europe ISO has a series of standards that cover the same material. The ASME and ISO standards use the same symbols, but interpretations are slightly different in some cases.

I

Influence Quantity A quantity that, in a direct measurement, does not affect the quantity that is measured, but affects the relation between the indication and the measurement result.

An example in laser tracker work would be air temperature and humidity.

Instrumental Drift Drift Change in a measured value, "related neither to a change in the quantity being measured nor to a change of any recognized influence quantity." (VIM)

Instrumental Uncertainty A component of measurement uncertainty arising from the measuring instrument or measuring system in use and obtained by its calibration.

L

Laser Scanning Using laser measurement systems to digitally capture the geometry of a product or part without contact between a probe and the part surface. The laser scanner will measure and record thousands or even millions of data points from a part's surface or a large area. Laser scanning is done on objects ranging in size from several inches to over 100 feet. These points, often called a point cloud, can then be used for comparison to customers' CAD models or to reverse engineer 3D models.

Laser Tracker One form of portable CMM, a laser tracker can measure large objects or spaces with a relatively high degree of accuracy. The typical laser tracker uses two rotary encoders and a laser ranging system to make measurements at the center of an SMR which is placed on the project or at the point of interest. Measurements are therefore limited by line of sight between tracker and SMR, and the tracker's range. Depending on the tracker and measurement conditions measurements can be made 100 yards or more from the tracker head.

A 3D laser tracker can be used in measuring, aligning, and even calibrating automated measurement systems, large and small parts, complex surfaces, and even immobile manufacturing machinery.

Least-Squares Adjustment A statistical method for providing a best-fit solution for measured [control] points by defining measurement error by minimizing the sum of the squares of measurement residuals. Where the residual is defined by the difference between the actual (measured or surveyed) value and the nominal (design) value typically defined by a 3D design model.

M

Maximum Permissible Error There are a few synonyms for this term, but all pertain to measurements and limits of error. The "extreme" value of measurement error, with respect to a known reference quantity value, permitted by specifications or regulations for a given measurement, measuring instrument, or measuring system.

NOTE 1 Usually, the term "maximum permissible errors" or "limits of error" is used where there are two extreme values.

NOTE 2 The term "tolerance" should not be used to designate "maximum permissible error." (VIM)

Measurement Result Any quantities attributed to a measurand from a measurement together with other relevant information. Environmental conditions or measurement uncertainty are two examples of other relevant information.

Measurement Traceability Also known as "metrological traceability." This is simply how metrologists connect to the standard by which they measure. That standard in the US is set and defined by NIST. Picture links in a chain. Each link represents a comparison that pertains to an instrument or tool's measurements to a NIST-defined standard or requirement. Because ECM calibrates to a NIST traceable standard, ECM can determine the precision and accuracy of a tool or instrument.

Measurement Uncertainty An estimate of the uncertainty of a measurement. Usually composed of instrumental uncertainty, and a number of other factors such as procedural uncertainty, and environmental uncertainty.

Measuring System A "set" of one or more measuring instruments and often other devices, including any reagent and supply, assembled, and adapted to give information used to generate measured quantity values within specified intervals for quantities of specified kinds.

NOTE 1 "A measuring system may consist of only one measuring instrument." (VIM)

Measurand Any quantity being measured. The term applies to any type of measurement, mass, force, luminosity, etc. in dimensional metrology, one could also call an object, or a feature of an object being measured a measurand and be understood.

Metrology The science of measurement and how measurements are used. Used to measure and verify an object's dimensional quality.

Mesh

1. A type of model consisting of small usually triangular planar surface patches that approximate an object's shape. It is much like an STL model. This type of model is often produced from scanned data and may be used as an intermediate step to a Nurbs surface model.
2. Occasionally used to describe Nurbs surfaces because the mathematical model of a surface can be thought of as a series of intersecting curves that lie in the surface.

Micron or Micrometer (μm) A unit in the metric system equal to one millionth of a meter or approximately 0.00003937 inches. It is commonly used to describe the uncertainty of precision measuring machines.

N

Noise At ECM, we use "Noise" to refer to incorrect data within a point cloud generated during a laser scanning. Reflections from dust in the air, errant reflections from corners, surface texture, or highly reflective surfaces are sources of noise.

Nominal Dimensions In dimensional metrology Nominal Dimensions are the dimension values given on a drawing or in a computer model. Measured values are compared to nominal to determine whether a part conforms to its design.

Noncontact Measurement This is simply the practices of taking an object's measurements without making physical contact with it. Often called scanning. Noncontact measurements can be used to measure an object with a delicate surface or weak structure that could not otherwise stand up to contact measuring.

Nonparametric Model Also known as a "dumb model," this is a 3D CAD model whose shape cannot usually be edited as easily as a parametric model. Commonly available file formats for porting files between CAD programs like IGES or STEP typically produce dumb models.

NURBS With a full name of Non-Uniform Rational B-Spline, this refers to NURBS curves and surfaces. The name comes from the mathematical technique used to model the curve or surface. It is one of the more common ways to model free-form geometry.

O

Outlier A point which lies outside the expected range of variation for data. In scan data outliers are usually removed during post-processing of the point cloud using statistical algorithms. Sometimes the scanning instrument and software remove them during the point cloud acquisition. In data generated by CMM or laser trackers, outliers are considered for removal on a case-by-case basis, because the data is often too sparse for statistical methods and because the operator can usually tell by visual observation or confirming measurements if an outlier accurately represents the part or was acquired incorrectly, perhaps by probing debris on the surface instead of the surface itself.

P

Parametric Model A parametric model is a CAD model that can be edited and changes will propagate through the model automatically preserving the relationships between features. Typically, these models can only be edited in the CAD program that created them.

Photogrammetry Noncontact imagery that takes 3D coordinate measurements (XYZ) through photographs.

Point Cloud A type of data consisting of many points in 3D space. The number of points may run from hundreds or thousands in the case of a CMM or laser tracker, to hundreds of millions, or even more points in the case of a laser scanner. Point cloud data can be compared to a CAD model or used to reverse engineer an object.

Polygonal Modeling A CAD model using small planar surfaces to approximate the shape of a surface. The surfaces are usually triangular, and the model is quite similar to an STL model. (See Mesh definition 1.)

Precision How close one measurement result will be to another result or set of results. Precision should not be mistaken for accuracy. A precise instrument could give a consistently erroneous result. The term "precision" is often used as a synonym for an instrument's repeatability.

Probe A device attached to a CMM that is used in taking measurements. Typical CMMs have contact probes (called touch probes) but a probe can be noncontact like a portable CMM laser scanner.

R

Reference Dimension A Reference Dimension is a dimension given to any features that is to be used only for reference or visualization purposes. These are called out on a drawing using parenthesis and do not get inspected.

Reference Model Typically a CAD model, but occasionally a physical master, against which data on a measured part is compared.

Repeatability A measurement system's precision under a set of measurement conditions, the repeatability condition.

Reproducibility Measurement precision under the Reproducibility Condition of Measurement.

Resolution

1. (of a measuring instrument or system) Smallest change in a quantity being measured that causes a perceptible change in the corresponding indication. Resolution can be affected by internal or external factors like noise, friction, or temperature. The value of the measurand may also affect resolution. Not to be confused with definition 2.

2. (of a displaying device) Smallest difference between indications that can be meaningfully distinguished or the number of digits in a digital display.

Reverse Engineering This process involves producing a drawing or 3D digital representation of a preexisting, tangible object usually via CAD-, CAE-, or CAM-type software. At ECM, the object is measured with a 3D laser scanner; the generated point cloud data is transferred into a NURBS surface or triangular mesh or a number of other options. From this point, the 3D digital CAD model is reconstructed to look like the original object that was laser scanned.

S

Scan Density Laser scan density refers to the proximity between 3D coordinates in a given data point cloud.

Scale Collectively transforming data to accommodate a change in temperature, units, or size.

Scan Speed This refers to a laser scanner's speed regarding the collection of 3D coordinates. The speed is measured in points per second or even millions of points per second.

Spatial Reference System (SRS) This is another name for a coordinate reference system (CRS) that captures geo-location entities. This would be considered a civil engineering term, or a term used in the utility field more so than in dimensional metrology.

Standard Deviation Standard deviation is a measure of the amount of variation in a set of numbers or values. A low standard deviation indicates that the values tend to be close to the mean (also called the expected value) of the set, while a high standard deviation indicates that the values are spread out over a wider range.

Standard Triangle Language (STL) STL is a file format native to the stereolithography, and metrology software created by 3D Systems. STL has several backronyms such as "Standard Triangle Language" and "Standard Tessellation Language." This file format is most often used for 3D printing and 3D scanning. STL files describe only the surface geometry of a three-dimensional object without any representation of color, texture, or other common CAD model attributes.

Stereolithography Stereolithography (SLA or SL) is a form of 3D printing technology used for creating 3D models, prototypes, and production parts.

T

Temperature Compensation This is something to consider when using a CMM. If using a CMM in less-than-ideal weather or within extreme temperatures, the CMM will swell or contract. To prevent in accuracy, this behavior needs to be considered when taking a 3D measurement unless the measurement system was built to prevent this and is highly accurate.

Thermal Compensation See Temperature Compensation and Coefficient of Thermal Expansion (CTE).

Tooling Ball A close tolerance sphere, usually on a precision shouldered shank used in aligning or locating parts or assemblies.

U

Uncertainty See Measurement Uncertainty.

V

Validation Demonstration by test or analysis that an instrument, system, or procedure is fit for a certain task.

Verification The proof, by means of objective evidence, that an item meets its specification requirements. In dimensional inspection the object's dimensions would be measured and compared to the drawing or CAD model of the object.

VIM The International Vocabulary of Metrology (VIM) is an international document that aims at disseminating scientific and technological knowledge about metrology by harmonizing the worldwide related fundamental terminology.

Index

Note: page numbers in *italics* refer to figures; those in **bold** to tables

---

Printed and bound by CPI Group (UK) Ltd, Croydon, CR0 4YY

27/10/2024

14580677-0001